Design of

LPG and LNG Jetties with Navigation and Risk Analysis

Liquefied petroleum gas (LPG) and liquefied natural gas (LNG) are both hazardous gases which are transported only in liquid form, under certain pressure, and/or at reduced temperatures. Though potentially hazardous, these gases can safely be transported by ships holding special types of containers which can be handled in jetty system specially constructed at seaports. This book covers design of the jetties used for LPG and LNG transportation, along with navigation and risk analysis.

Besides the designing of the jetty, fendering system, navigational channels, tanker moorings, requirement of shipping tugs, etc., also discussed is hazard analysis along with rules and regulations specific to handling LPG and LNG tankers in a port complex.

The book is expected to serve as a handy guide and reference to all the engineering personnel engaged in designing of LPG and LNG jetties including navigation and administering the seaports and navigation channels keeping the operational hazards in check.

Design of

LPG and LNG Jetties with Navigation and Risk Analysis

LN PATNAIK
BE (Civil), DIIT (Bombay), MIE, CE (India)
Consulting Engineer

Ex-Chief Engineer
Inland Waterways Authority of India (IWAI)
Ministry of Shipping, Government of India
New Delhi

CBS Publishers & Distributors Pvt Ltd
New Delhi ▪ Bengaluru ▪ Pune ▪ Kochi ▪ Chennai

Design of

LPG and LNG Jetties with Navigation and Risk Analysis

ISBN: 978-81-239-2030-6

First Edition : 2011

Published by Satish Kumar Jain and produced by Vinod K. Jain for

CBS Publishers & Distributors Pvt Ltd

4819/XI Prahlad Street, 24 Ansari Road, Daryaganj
New Delhi 110 002, India.   Website: www.cbspd.com

Ph: 23289259, 23266861, 23266867   e-mail: delhi@cbspd.com
Fax: 011-23243014   cbspubs@vsnl.com
                                          cbspubs@airtelmail.in.

Branches

- Bengaluru: Seema House 2975, 17th Cross, K.R. Road,
  Banasankari 2nd Stage, Bengaluru 560 070, Karnataka
  Ph: +91-80-26771678/79   Fax: +91-80-26771680   e-mail: bangalore@cbspd.com

- Pune: Bhuruk Prestige, Sr. No. 52/12/2+1+3/2 Narhe, Haveli
  (Near Katraj-Dehu Road Bypass), Pune 411 051, Maharashtra
  Ph: 020-64704058, 64704059, 32342277 Fax: +91-020-24300160 e-mail: pune@cbspd.com

- Kochi: 36/14 Kalluvilakam, Lissie Hospital Road, Kochi 682 018,
  Kerala
  Ph: +91-484-4059061-65   Fax: +91-484-4059065   e-mail: cochin@cbspd.com

- Chennai: 20, West Park Road, Shenoy Nagar, Chennai 600 030,
  Tamil Nadu
  Ph: +91-44-26260666, 26208620   Fax: +91-44-45530020   email: chennai@cbspd.com

Printed at India Binding House, Noida, UP

Preface

It was while designing the LPG jetty of Gujarat Pipavav Port including the risk analysis and obtaining the approval of Chief Controller of Explosives, Nagpur, as required under the relevant rules, I felt the necessity of a comprehensive book dealing with all the aspects for handling LPG/LNG tankers in a jetty system through a seaport such as the design of the jetty, fendering system, navigational channels, tanker moorings, requirement of shipping tugs and hazard analysis, and rules and regulations specific to handle LPG/LNG tankers in a port complex. It is hoped that this book would be useful for all those who are engaged in dealing with the design of LPG/LNG jetty system in a seaport.

This book will also be applicable for design of jetties for handling crude POL and chemical gas tankers.

I express my gratitude to several authors and institutions from which several figures and materials have been derived and incorporated in the book particularly to McGuire and White for their excellent book *Liquefied Gas Handling Principles on Ships and Terminals*, Carl Theoresen for his book *Port Design—Guidelines and Recommendations*, Hans Agerschou et al *Planning and Design of Ports and Marine Terminals*, Per Bruun for his book *Port Engineering*, Centre for Advanced Maritime Studies, Edinburgh, for their safety course on *Liquefied Gas Carriers*, PF Willerton for the book *Basic Ship Handling*, Capt. Malcolm Armstrong for *Practical Ship Handling*, and several others acknowledged in the References. I also wish my appreciation to the British Standards Institution and Indian Standards Institution for using their relevant code of practice, and to Rane Elastomer Processors, Bombay, and Sibata Fenders, Japan, for using their company brochures on fenders.

I also wish to express my appreciation to Capt. D Chaudhury, Ex-Deputy Conservator (DC), Paradip Port Trust (PPT); Capt. BS Kumar, Ex-Harbour Master, PPT (now Advisor, Tatas); Capt. UC Patnaik, Ex-DC, PPT; and Capt. PBS Patnaik, Ex-DC, Vizag Port, for their guidance while designing the navigational channels in general.

Gratitude is also expressed to late Mr SBN Singh, Vice-Admiral (Rtd), Commodore (Rtd); VG Honnaver and Mr D Dikshit, Technical Advisor, Gujarat Pipavav Port, for the help given to me while designing the LPG jetty of Gujarat Pipavav Port.

Most of the subject chapters dealt in this book have been published in *Indian Ports*, a quarterly journal published by the Indian Ports Association, New Delhi, a body sponsored by all the major ports of India that are under the Ministry of Surface Transport, Government of India, New Delhi.

Helpful criticism is invited which would be gratefully received and would be incorporated in subsequent editions.

LN Patnaik

uniconpune@rediffmail.com

Contents

3. DESIGN OF JETTY FENDERING SYSTEM   22

4. DESIGN OF NAVIGATIONAL CHANNELS   39

List of Figures

Brief Description of LPG and LNG Tankers Now in Service

1.1 INTRODUCTION

LPG is abbreviation for liquefied petroleum gas and LNG for liquefied natural gas. Both are hazardous gases and are transported at atmospheric pressure only after converting them into a liquid form either under certain pressure or reducing the temperature.

LPG comprises propane, butane and a mixture of these two. LNG comprises predominantly methane. The physical properties of these gases are given in Table 1.1.

Gas	Atmospheric boiling point-z)	Liquid relative density at atm. Boiling Point (water=1)	Vapour relative density (Air=1)
Methane	−161.5	0.427	0.554
Propane	−42.3	0.583	1.56
η-Butane	−0.5	0.600	2.09
i-Butane	−11.70	0.596	2.07

Table 1.1: Physical properties of gases

LNG is produced from the natural gas after removing ethane Natural Gas Liquids (NGL) pentane (C_5), heavier fractions, water, carbon dioxide, nitrogen and other non-hydrocarbon contaminates as shown in Fig. 1.1. Natural gas is transported either by a pipeline as a gas or by sea in its liquefied form as LNG. LPG is extracted either from natural gas or from refining crude oil and when produced in this way, they are usually manufactured in pressurised form.

At a pressure of 7 bars or at a temperature of (−) 46°C, LPG is liquefied and can be transported at atmospheric temperature. Likewise at a temperature of (−) 161°C LNG, is liquefied and can only be transported at the atmospheric

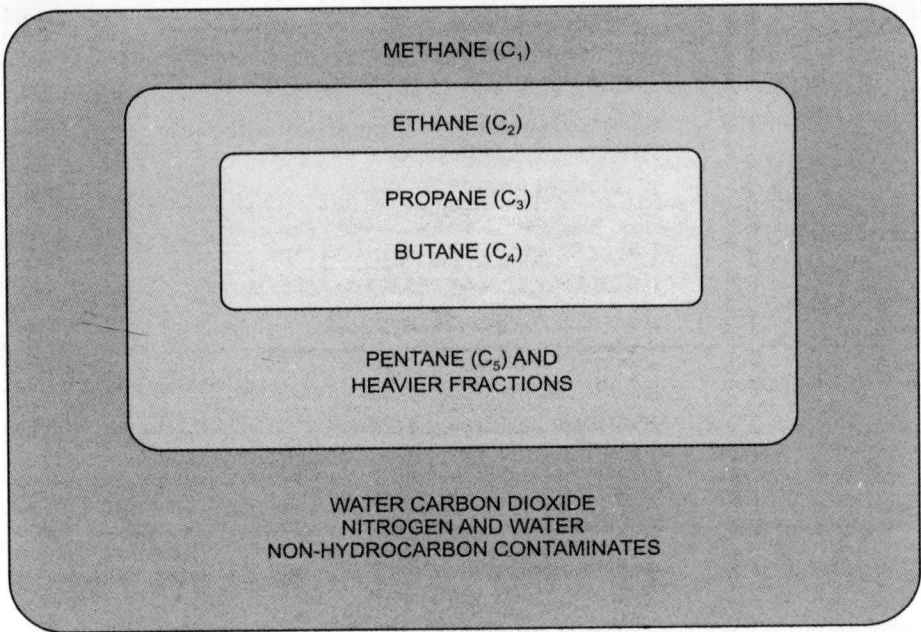

Fig. 1.1: Constituents of natural gas

temperature. LPG and LNG are potentially hazardous. However, these gases can safely be transported by ships holding special types of containers.

1.2 GAS CARRIERS

1.2.1 LPG Carriers

LPG is transported using the following category of tankers:

- Fully pressurised
- Semi pressurised
- Fully refrigerated

1.2.2 Fully Pressurised Ships

The containment system in most of the fully pressurised carriers comprises of two to three horizontal, cylindrical or spherical cargo tanks fabricated out of carbon steel and carry their cargoes at ambient temperature with tanks designed for 18 to 20 bar pressure. These pressure ranges necessities thicker wall plating. Consequently the tanks which may be Horton Spheres, are extremely heavy; cargo capacity varying between 4000 m³ and about 10,000 m³. Ballast is carried in double bottom and in top wing tanks. No thermal insulation or re-liquification plant is necessary. Fully pressurised ships are the simplest of all gas carriers. These ships can carry LPG and also ammonia and are generally deployed in short sea trades.

1.2.3 Semi-Pressurised Ships

These category of ships are similar to the fully pressurised ships but the containment is designed for a maximum working pressure of 5 to 7 bar and to a temperature of (–) 48° C at which range most of the LPG and chemical gas cargo can be carried. These tanks are usually made from low temperature steels. A complete double hull is required for these category of ships.

1.2.4 Fully Refrigerated Ships

The cargo containment system comprises of the following:
- Independent tanks with single hull but double bottom and hopper tanks.
- Independent tanks with double hall.
- Integral tanks (incorporating a double hull).
- Semi-membrane tanks (incorporating a double hulls).

The cargo is carried approximately at atmospheric pressure and can also carry LPG/Ammonia. The tanks are constructed out of low temperature steel to permit carriage temperature of –48°C and for a maximum working pressure of 0.7 bar. These category of ships ranges from 20,000 to 100,000 m³.

1.3 LNG CARRIERS

The containment system of LNG carriers are now mainly of four types:
- Gaz Transport membrane system.
- Technigaz membrane system.
- Kvaerner Moss special independent type system.
- IHI SPB Tank prismatic system.

All LNG carriers have double hulls throughout their cargo length which provide adequate space for ballast. Hold space around to cargo tanks are continuously inerted except in case of spherical type 'B' containment where hold spaces are filled with dry air provided that there is an adequate means for inerting such spaces in the event of cargo leakage. Continuous gas monitoring of all hold spaces is required.

These carriers are now built to transport large volumes of LNG (1,25,000 m³ to 1,35,000 m³) at its atmospheric boiling point of about (–) 162° C. All boil-off gas is burnt as fuel in ships main boiler. In Gaz membrane, the primary barrier is a stainless steel alloy containing about 36% nickel and 0.2% carbon. In Technigaz membrane system the primary barriers is of stainless steel as shown in Figs 1.2 to 1.5.

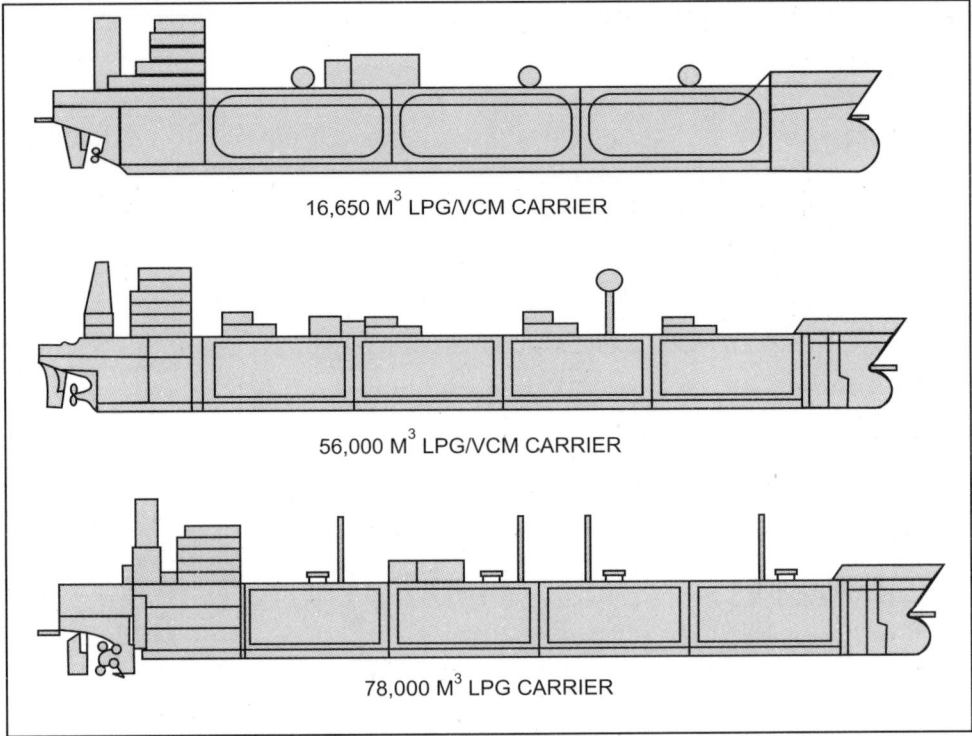

16,650 M^3 LPG/VCM CARRIER

56,000 M^3 LPG/VCM CARRIER

78,000 M^3 LPG CARRIER

Fig.1.2: LPG carriers (elevation)

1.4 DESIGN SHIP SIZE

An LPG or an LNG jetty should be designed to accommodate a maximum and minimum range of tankers proposed to be serviced. The maximum size should be based on a ship trend analysis, taking account of the depths at the proposed harbour that can be obtained economically.

The fleet composition of gas tankers in 1994 was of the following order:

LNG carriers	:	85
Fully refrigerated ships	:	165
Ethylene carriers	:	100
Semi-pressurised ships	:	200
Pressurised ships	:	350

LNG carriers currently deployed are in the range of 1,25,000 to 1,35,000 m^3. Fully refrigerated ships for LPG range between 20,000 to 1,00,000 m^3. Ethylene carriers range between 1,000 to 12,000 m^3. Fully pressurised ships for LPG

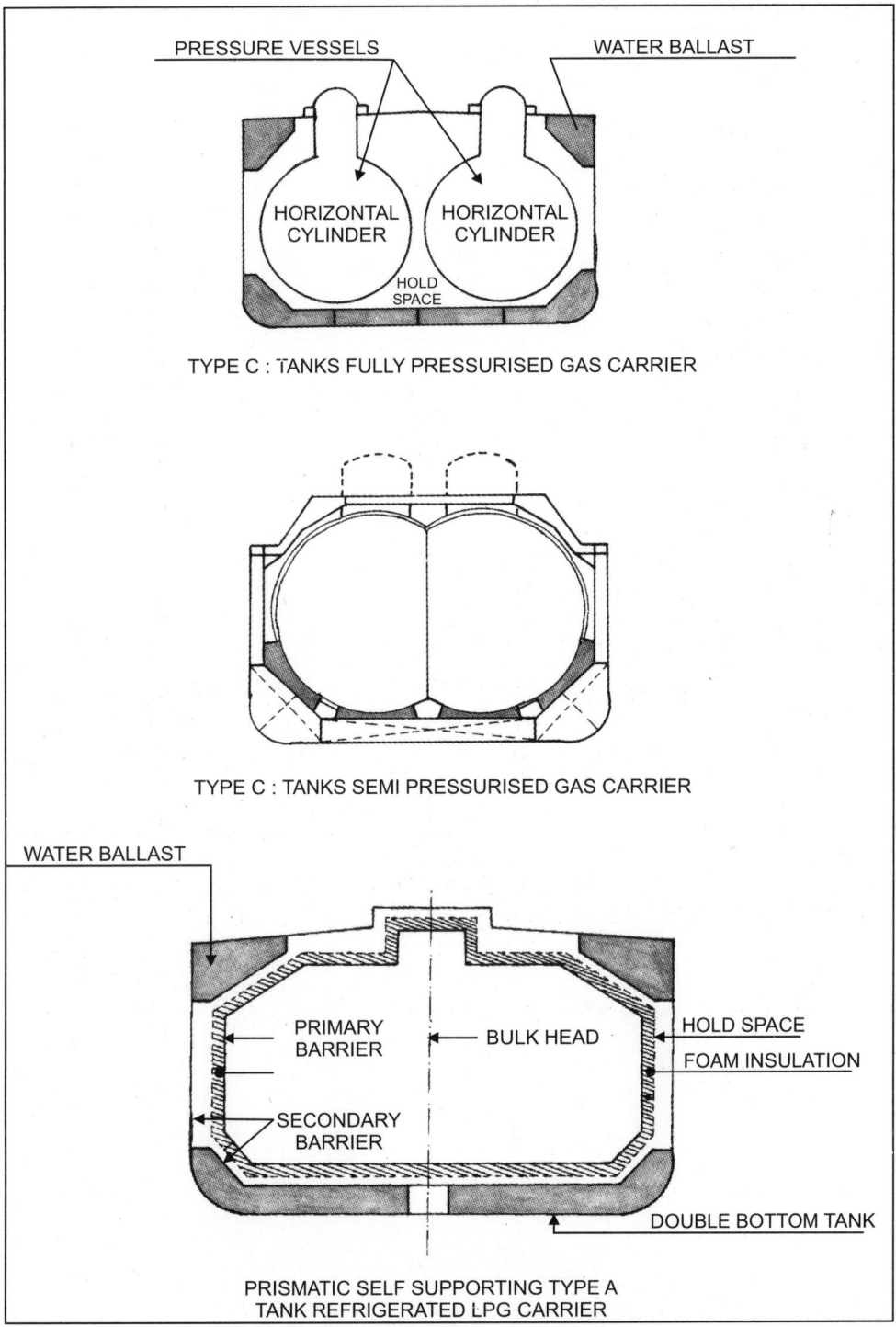

Fig.1.3: Typical cross-sections of larger LPG carriers

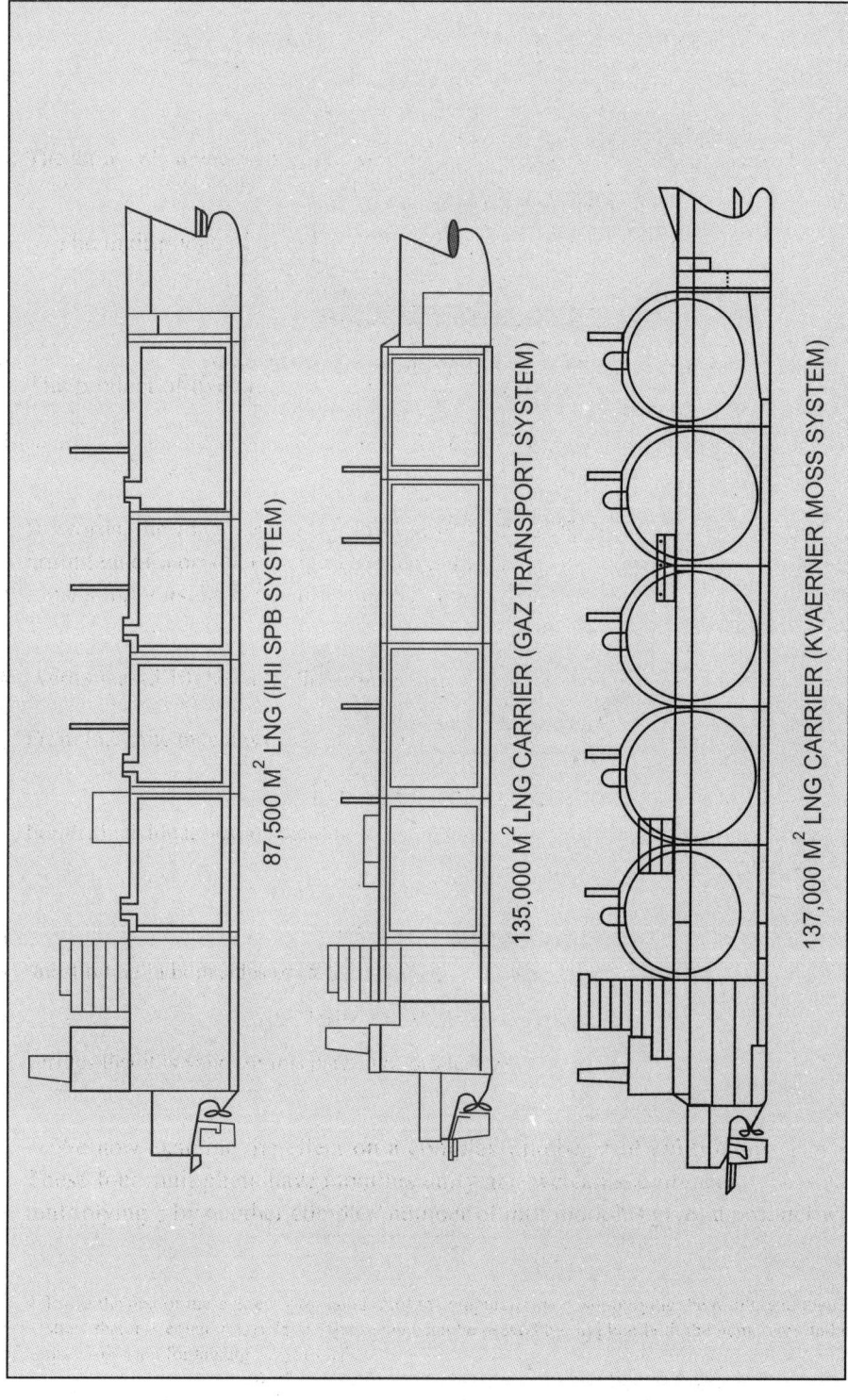

87,500 M^2 LNG (IHI SPB SYSTEM)

135,000 M^2 LNG CARRIER (GAZ TRANSPORT SYSTEM)

137,000 M^2 LNG CARRIER (KVAERNER MOSS SYSTEM)

Fig. 1.4: LNG carriers (elevation)

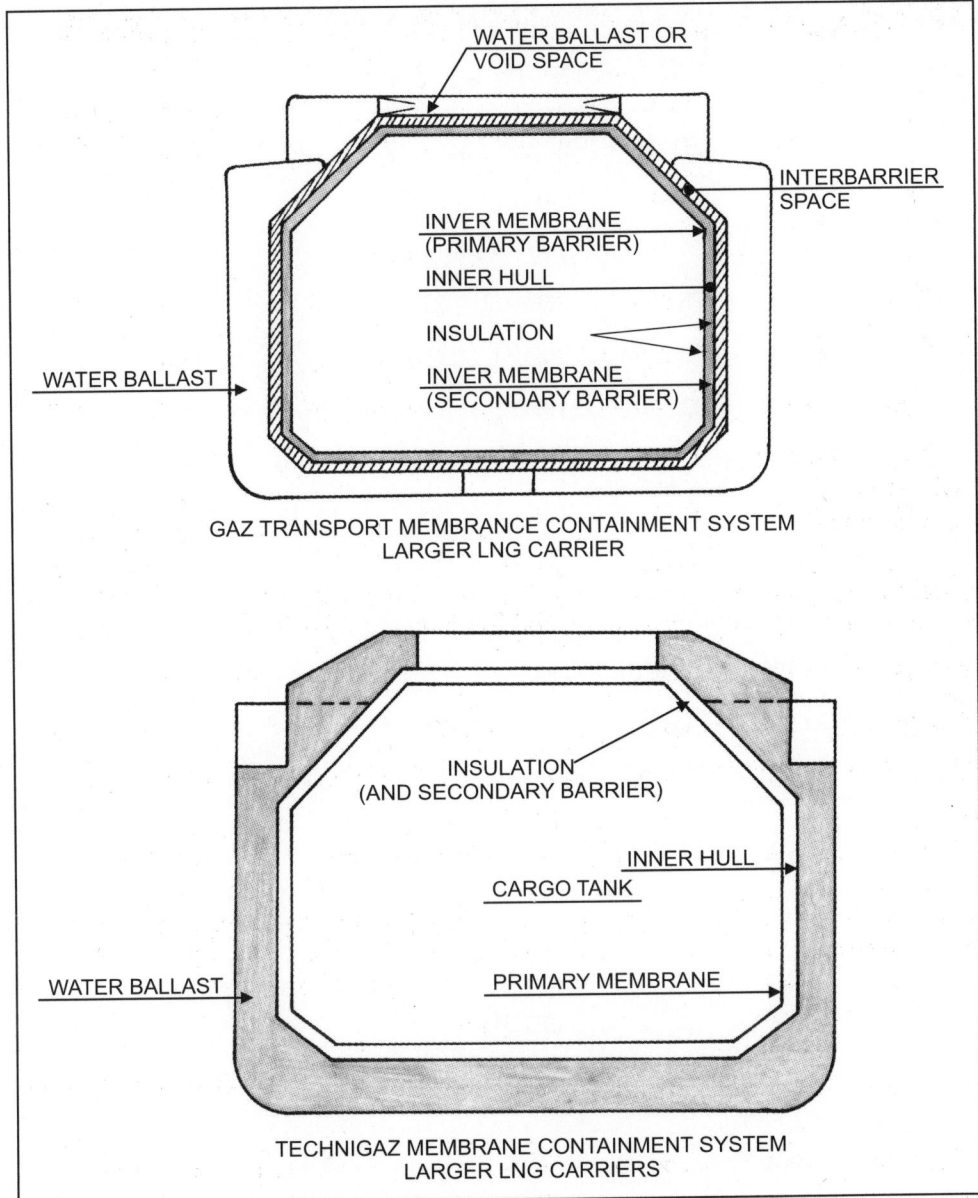

Fig.1.5: Typical cross-sections of LNG carriers

range between 4,000 and 10,000 m³. In longer hauls, LNG carriers and fully refrigerated ships above 70,000 m³ are generally deployed. Table 1.2 shows particulars of the LPG/LNG carriers.

Table 1.2: Particulars of LPG and LNG tankers

Capacity	DWT	Displacement	Over all length	Length between perps	Beam	Moulded depth	Draft max
m³	ton	ton	m	m	m	m	m
LPG Tankers							
75,000	46,900	75,000	229	218	36.0	21.0	12.1
52,000	38,500	53,200	206	196	31.4	18.6	11.3
24,000	18,100	32,800	157	149	25.3	16.0	10.1
15,000	16,200	25,000	151	140	25.0	14.3	9.6
5,000	5,400	9,000	106	98	17.0	10.0	7.4
2,500	2,800	5,000	75	70	14.0	7.9	6.8
LNG Tankers							
1,25,000	73,100	1,02,000	294	281	41.6	25.0	11.7
87,600	53,600	74,000	250	237	40.0	23.0	10.6
29,000	22,100	32,600	182	171	29.0	16.5	9.0

2

Physical Planning of Jetty System for Handling LPG and LNG Tankers

2.1 SITE SELECTION

LPG/LNG are hazardous cargo and are associated with fire, the intensity of radiated heat from the flame, the blast over pressure in the event of an explosion would cause extensive damage and destruction to the personnel and properties as enumerated in the succeeding Chapter 7 (Hazard Analysis). Owing to the hazards involved, the location of an LPG/LNG jetty site in a port complex is an important matter. The site should be tranquil enough for safe working of the tankers. The jetty needs to be placed as parallel as possible to the predominant wind and current directions so as to prevent strong cross forces. The surge, sway and roll motions should remain within the tolerable limits. The wave height at the jetty site should also remain about 30 cm under severe wind and wave conditions. This, however can be assessed after conducting a model test (mathematical/physical).

The jetty and the tank farm should remain at least 1500 m away from the places of human concentration like Schools, Colleges, Market places, Temples, Masjids, Hospitals, etc.

The fire fighting pump house should also remain outside of LFL distance or 1 psi blast over pressure zone whichever greater.

The navigation of LPG/LNG tankers should be restricted to day light hours only and while these gas tankers are in transit, movement of other vessels are to be stopped till the gas tanker is made to berth alongside of the designated jetty. The tanker should invariable be assisted by tugs, pilot on board the tanker.

The site should be free of penetration of long period waves (T>14 sec.) and should also be free from the rock strata otherwise rock dredging would be involved, which would prove to be very expensive.

While docking an LPG/LNG tanker alongside a jetty, parallel berthing system should be followed limiting the angle of approach to 3° or at the most to 5° with the assistance of tugs. Stern in first so that the bow faces deep waters. In event, the tanker catchers fire or there is a fire on board the jetty, the tanker should be pulled off the jetty and taken to roads.

2.2 COMPONENTS OF THE JETTY

A typical LPG/LNG handling project would comprise of the following:

- LPG/LNG jetty system.
- Approach bridge with a roadway and various pipe rakes which would connect the jetty with the shore.
- Fire water intake jetty and pump house.
- Tank farm for storing LPG/LNG products with loading and unloading arrangements including the fire fighting system.

2.2.1 LPG/LNG Jetty System

The jetty system would comprise of the following:

- A Central Unloading Platform (CUP)
- Two or more Breasting Dolphins on either of the CUP
- A set of Mooring Dolphins
- An approach bridge containing a roadway for vehicular access to the CUP, a pipe rake for the LPG/LNG pipelines, a separate pipe rake for the fire fighting pipelines, fresh water pipe line and a cable tray. This approach bridge will connect the jetty with the shore.

2.2.2 Central Unloading Platform (CUP)

The Central Unloading Platform (CUP) will accommodate the metallic unloading arms, LPG/LNG pipelines, which would be connected to the tank farm that would contain the storage tanks, upstream of the jetty, fire fighting pipelines, fresh water pipeline, electric cabling system and a control tower. The CUP also would carry two spring lines of 20 T capacity near to the berthing face of the jetty.

2.2.3 Breasting Dolphins (BD)

On either side of the CUP, two breasting dolphins would be provided, equipped with fenders to take the impact of the berthing tankers. To ensure contact on the parallel side of the tanker, the breasting dolphins should be set apart by 0.25 to 0.40 x LOA of the tanker. The breasting dolphins are generally detached from the CUP so as to protect it against any damage that may occur during the berthing operation of the tanker. Two fire water monitor towers in each of the breasting dolphins should be provided for fire fighting purposes.

2.2.4 Mooring Dolphins (MD)

On either side of the breasting dolphins 2 to 3 sets of mooring dolphins are provided to take the mooring ropes of the tanker as shown in Fig. 2.1. The mooring dolphins are placed about 35–50 m off the berthing face of the jetty.

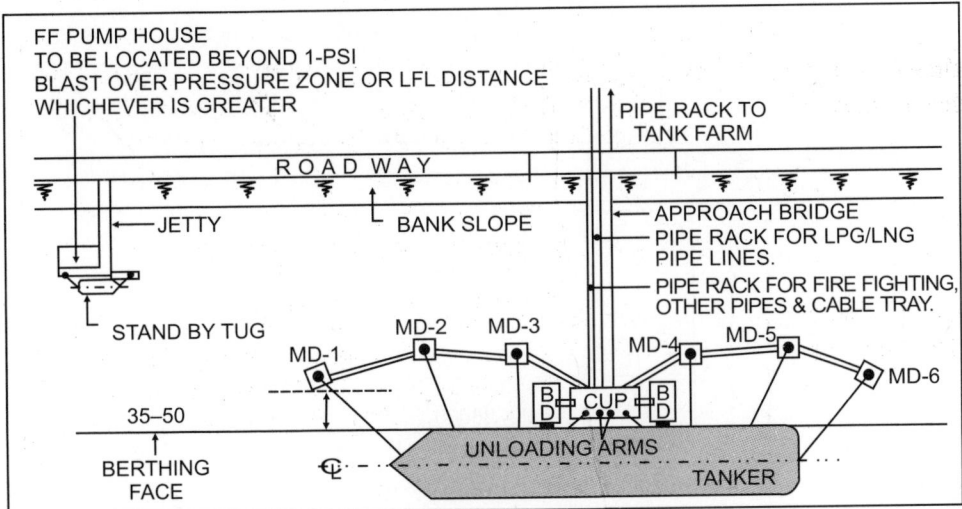

Fig. 2.1: Plan of an LPG/LNG jetty system

If the jetty is required to service one set of tankers, then two breasting dolphins on either of the sides of the CUP will be sufficient. But if the jetty is required to handle a large category of tankers, small, medium and large, 4-breasting dolphins need to be provided as shown in Fig. 2.2. That apart, one set of additional mooring dolphins may have to be provided.

2.2.5 Approach Bridge

An approach bridge would be provided which will connect the CUP with the shore. It would have a road way and also the pipe rakes as described earlier. A typical cross-section of this bridge is shown in Fig. 2.3.

2.2.6 Cargo Transfer System

LPG/LNG are transferred from a tanker berthed alongside of a jetty though a metallic hard arm (unloading arm). The arm is fitted with swivel joints to provide the required movement between the tanker and the jetty connections, maximum angular limits are typically 150° at the apex angle, 15° on the arm tuck back from the vertical, 10° for arm elevation above the horizontal and about 45° for slew or luff. Figure 2.4 shows a typical unloading arm.

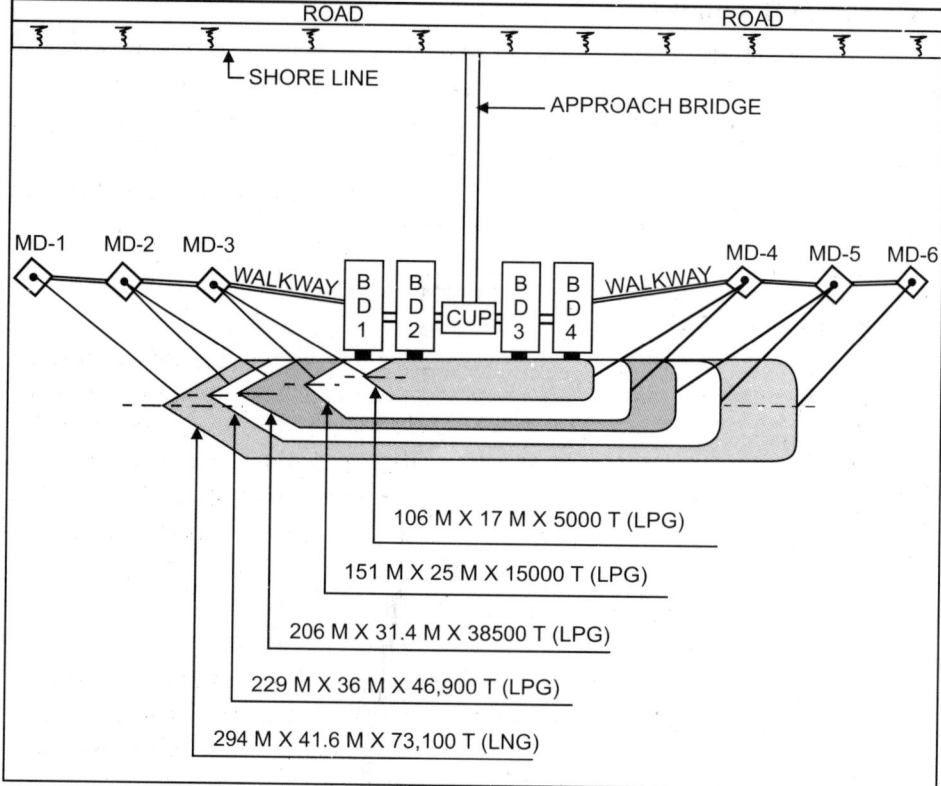

Fig. 2.2: Plan of a multiuse LPG/LNG jetty system

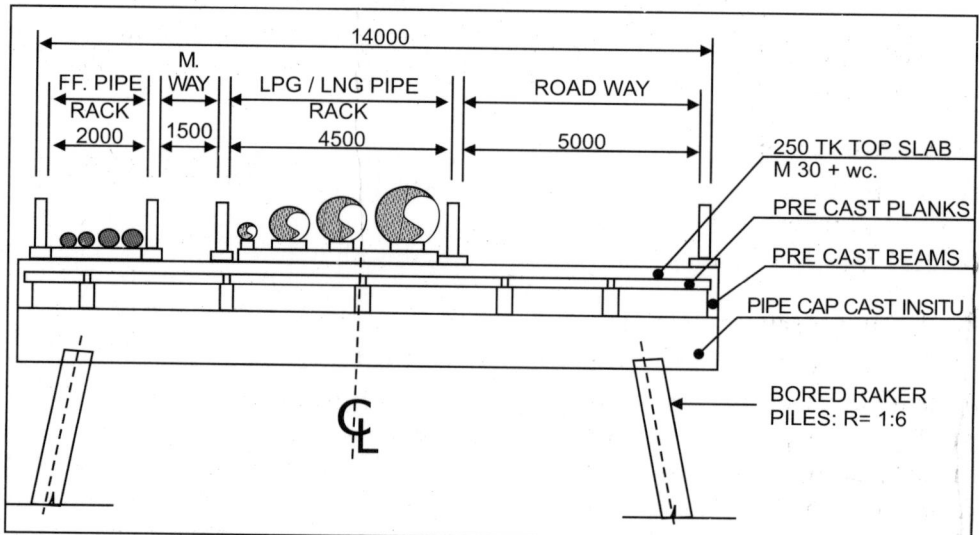

Fig. 2.3: Typical section of an approach jetty

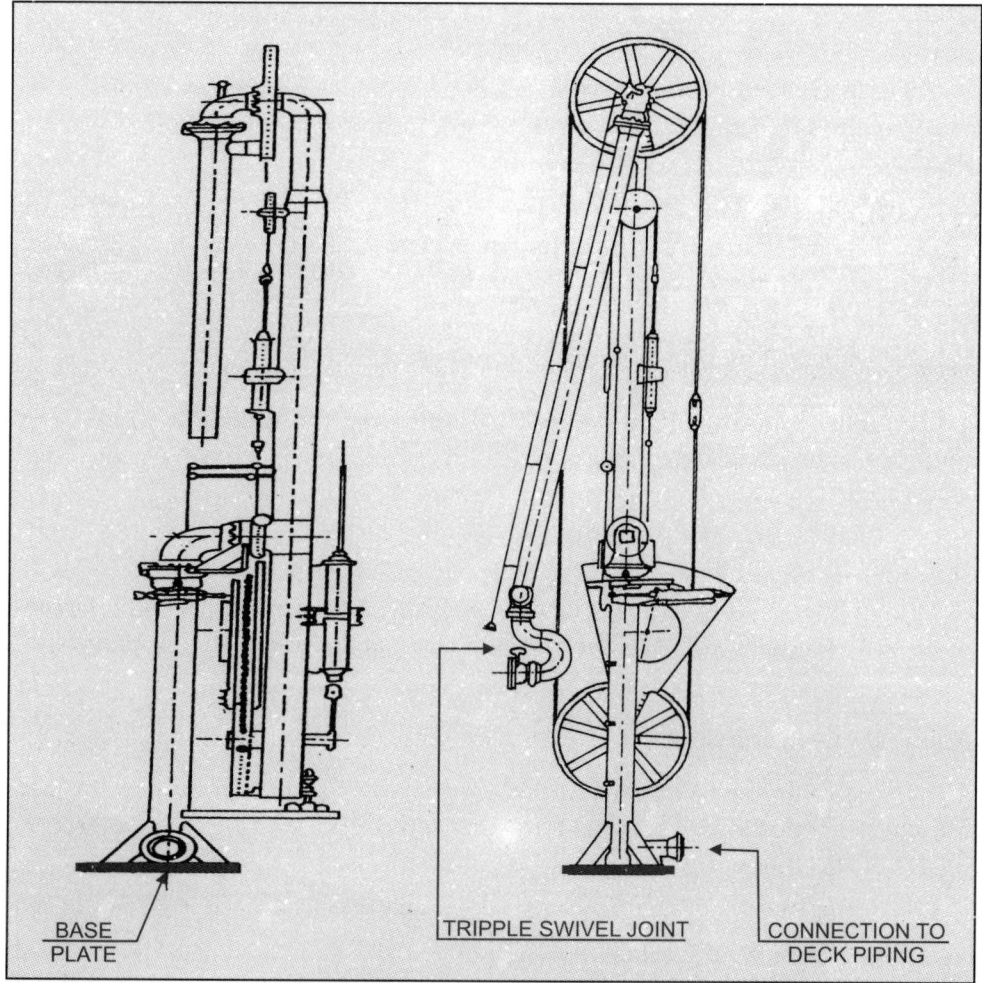

Fig. 2.4: Typical marine hard arm

The connecting system between hard arm and the tanker can be bolted flanges or a quick connect/disconnect coupling (OCDC). Should the limits of the design operating envelope are approached or some other emergency occurs, the emergency disconnect arrangements can be used.

2.2.7 Hoses

Cargo transfer can also be accomplished by using hoses. These hoses comply with the British Standard: BS - 4084—Hoses and Hose Assemblies for LPG. These are three type of cargo hoses suitable for the liquefied gases. They can be of composite rubber or stainless steel construction. For LNG they can be of composite construction or of corrugated stainless steel, but in general the composite type is preferred. LPG hoses may be of similar construction to those used for LNG but hoses of synthetic rubber manufacture may also be used.

These hoses are correctly supported in a hose handling cradle so that manufacture's recommendations on the minimum bending radius are maintained.

2.2.8 Design Loadings

Live load intensity in an LPG/LNG jetty is not vary significant. However, following loadings should be considered:

2.2.8.1 Central Unloading Platform

The Central Unloading Platform (CUP) will carry the unloading arms, self weight of which will be supplied by the manufacturer. That apart, the CUP will carry LPG/LNG pipelines, other pipelines, a central observatory tower and two spring lines near to the berthing face of the jetty with $10°$ inclination to the horizontal. As such, the CUP should be designed for the following loadings:

(a) D.L + 2 Nos. or more unloading arms + UDL of 1.5 T/m^2 which will take care of other loadings + 2-20T line pull with $10°$ inclination to the horizontal.

(b) Seismic load as per 1S – 1893 (to be considered separately).

2.2.8.2 Mooring Dolphins

Mooring dolphins should be designed for the line pull depending on the size of tankers proposed to be serviced at the jetty. The loadings will therefore consist of the following:

(a) DL + line pull from the mooring ropes inclined at an angle of $30°$ to $60°$ + LL of ½ T/m^2

(b) Seismic load as specified above.

2.2.8.3 Breasting Dolphins

Breasting dolphins should be designed for the following loadings:

(c) DL + Berthing Loads + water monitor tower for fire fighting purposes for which an UDL of 1.5 T/m^2 will prove sufficient.

(d) Seismic load as specified above.

2.2.8.4 Approach jetty

The approach jetty would carry a single lane roadway to carry-out routine maintenance of the jetty and would accommodate the LPG/LNG pipe rack, fire fighting pipelines, fresh water pipelines and a cable tray. As such, the approach jetty should be designed for the following loadings:

(a) D.L. + I RC – Class-B loading for the roadway + 1.5 T/m² for the pipe racks.

(b) Seismic load as specified above.

2.3 BOLLARDS AND QUICK RELEASE HOOKS (QRH)

Bollards or quick release hooks should be provided so as to tie down the tanker effectively against the environmental forces due to wind, current and wave that would be prevailing at the site of the jetty. The bollard pulls for different category of tankers as per 15-4651-P.III, are as under:

Table 2.1: Bollard pulls	
Displacement (Tons)	Line Pull (Tons)
2,000	10
5,000	20
10,000	30
20,000	60
50,000	80
1,00,000	100
2,00,000	150

For locations of exceptional wind, current and the other adverse forces, these mooring point loads should be increased by 25% (B.S. 6349 P-4).

2.3.1 Bollards

Generally speaking, bollard casings could be of cast steel. They may be of pillar type, tee head type, twin head type and sloping lobes type. In several port of India, mild steel bollard posts have also been used.

2.3.2 Bollard Design

As per 1S – 4651 (P-III) pull angle up to 30° above the horizontal should be considered. Some authors for the purpose of design, take this angle as 60°. As such bollards should be designed for angles of 30° to 60° inclination.

2.3.3 Quick Release Hook (QRH)

Instead of the usual bollards, the mooring dolphins of an LPG/LNG jetty should be equipped with QRH. These hooks are available in standard designs with single, double or triple hook assemblies. The unit with triple hook assembly should however be adopted. The hook assembly shall support design loads from 10° below horizontal to 30° above. These hooks shall be electrically operated either locally or remotely operated from the control tower.

These shall be of being released swiftly and effectively at all conditions, viz., being slack, tensioned to the hooks, design load or an intermediate load. The system is to be designed to prevent accidental release of the hooks.

The mooring hooks shall be of cast or forged steel. These shall be intrinsically safe with energy absorbing pads to prevent sparking and damage to any part of the assembly. After releasing, the hook should return to its original position without aid.

2.3.4 Powered Capstans

Each QRH assembly will have a powered capstan mounted over it to pull mooring lines to any mooring hooks on its assembly by means of fibre messenger ropes. The capstan shall be designed for a line pull of 2000 kgs at a normal operating lines speed of 25 m/min. It shall be motor driven with a motor rating of a 7.5 kW operating on 433 V. The capstan shall have a 'deadman-type' foot-operated pedal for "on-off-reverse switch". It shall also have a bidirectional brake which can freeze the rotation unless the foot-pedal is depressed.

The capstans shall be fixed in such a way that the messenger rope being pulled by it shall not be higher than 1.2 m from the dolphin's working deck.

2.4 FIRE FIGHTING FACILITIES

LPG and LNG are being hazardous cargo, comprehensive fire fighting facilities, based on OISD norms, covering the jetty and the pumping terminal should be provided in consultation with a specialised agency. Sea water can be used for fire fighting purposes. In gist, the following facilities are to be provided:

(1) Fire water system which would broadly comprise of the following components:
   (a) Fire water pump house
   (b) Tower mounted water monitors on either side of CUP/breasting dolphins
   (c) Nozzles for the water curtain along the seaside face of the CUP; hydrants and monitors located at intervals on the deck and the pumping terminal.
(2) Helon protection system shall be provided for the control room and in the pumping terminal.
(3) Mobile fire fighting equipment to be provided.
(4) First aid fire-fighting equipment to be provided.
(5) Fire detection and alarm system to be installed at specific locations.

2.4.1 Fire Water Intake Jetty

A fire water intake jetty with the pump house shall be provided. This jetty will be located outside of LFL or 1 psi blasé-over pressure zones whichever is greater as shown in Fig. 2.1.

2.4.2 Standby Tug

A standby tug should be retained near the jetty during LPG/LNG operations so that the jetty can be evacuated by pulling off the tanker at the shortest possible time, in case of an emergency and taken to roads, as shown in Fig. 2.1.

2.4.3. Fire Detection and Alarm System

The following areas shall be provided with gas detectors:
- Service Platform (CUP) along with pipe manifold and valves.
- Unloading arms.
- Pumping terminal.

2.5 DESIGN OF AN LPG / LNG JETTY

Generally speaking, an LPG/LNG jetty is a open jetty structure and is first constructed at the designated site and later connected to the navigational channel of the port by capital dredging. As would be evident from the succeeding para, the jetty would essentially consists of the following:
- Breasting dolphins to take the impact of the berthing tankers.
- Mooring dolphins to take mooring ropes of the berthing tankers.
- A central unloading platform to carry unloading arms and pipelines.
- An approach bridge with a roadway and pipe rakes.

Majority of these structures will be subjected to loads in form of horizontal force which would cause bending in the above elements. As such, predominantly the bending moment on piles of the breasting dolphins and mooring dolphins would be quite prominate as compared to the vertical reactions. Accordingly, these elements should be designated as bending elements.

2.6 PILE GROUP ANALYSIS

The dolphins and the CUP may consists of only vertical piles for easy construction. Here the piles would be subjected to a large amount of BM with a small amount of the vertical reaction. Consequently the structure may relatively be costlier. However, as an alternative, the bend may consist of not only the vertical piles, but also Raker piles (batter piles), raking in two directions. This will make the pile bend more rigid and would reduce the BM and would increase the vertical reaction in piles. Nevertheless, the bend

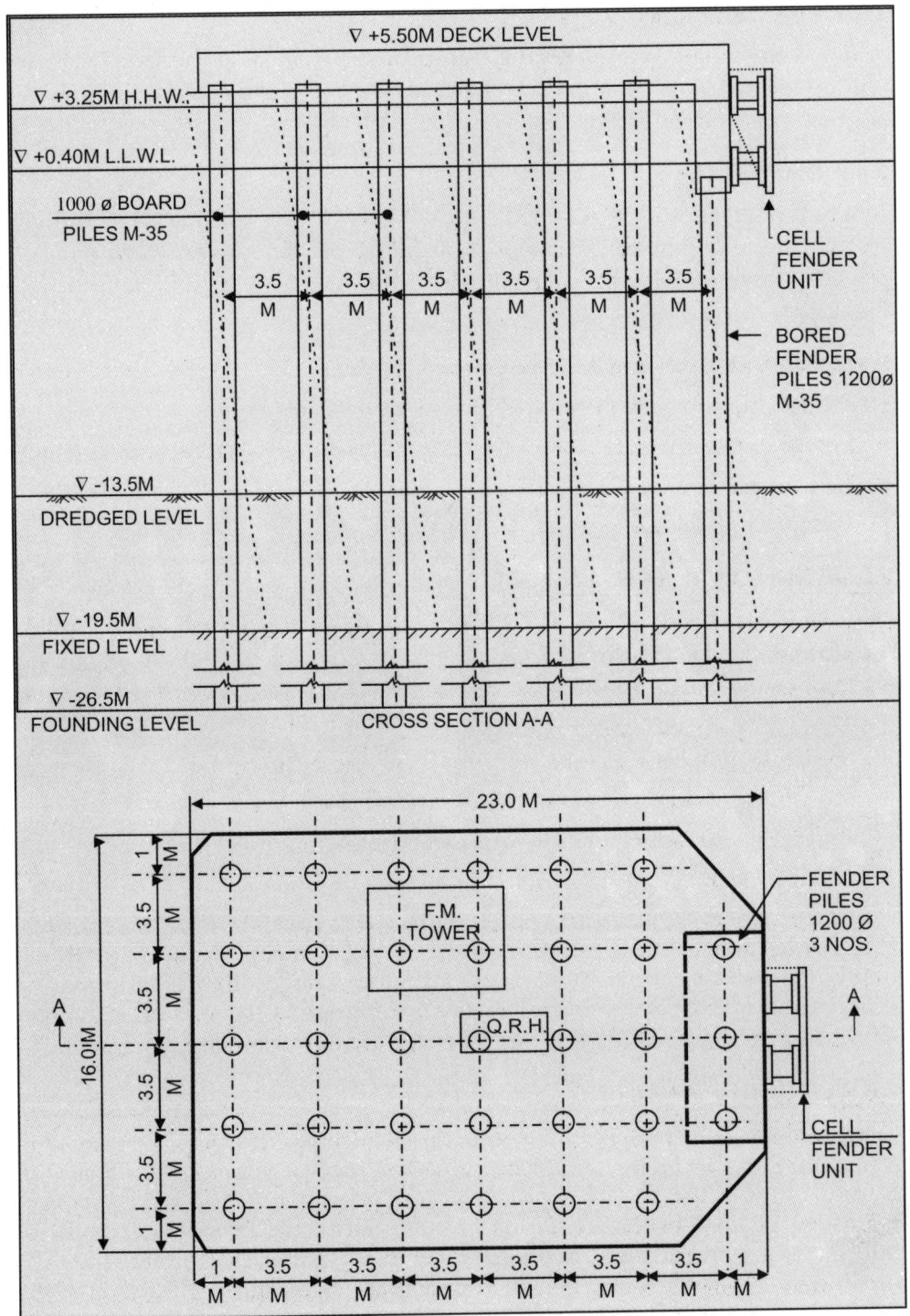

Fig. 2.5: Plan of breasting dolphin

will be a statically inderminate structure and pile load analysis shall have to carried out using a software STAD-III which may give more accurate results. However, under certain assumptions, a simple statistical method of analysis can be made which will give a quick solution. This has been illustrated in the following para.

2.7 DESIGN EXAMPLE

Design the breasting dolphin piles, fendering system of which has been designed and given in Chapter 3, particulars of which are reproduced below.

Capacity (T)	DWT (T)	LOA (M)	BEAM (M)	DRAFT (M)	LPP (M)
75,000	46,900	229	36	12.1	218

A typical plan along with a cross-section of the breasting dolphin are given in Fig. 2.5. As assessed in Chapter 3, reaction on the breasting dolphin in the transverse direction is 300 T. Take 20% as the longitudinal component of the force, i.e., 60 T. As the pile section is circular with symmetrical reinforcement, both the reactions can be added up and the analysis can be made. Hence, maximum reaction due to impact of berthing tankers would be 360 T. On this basis and taking into account the following data, pile load analysis has been made:

- Impact force = 360 T
- Cut off level of piles = + 3.85 M
- Dredge level = (–) 13.5 M
- "Fixed" level of piles below the dredged level = 13.5 + 6 × 1 (as the soil is sandy clay) as per 1S.4651.......... (–) 19.5 M
- Founding level of piles = (–) 26.5 M

- Point of contra flexure = $\dfrac{19.5 + 3.85}{2} = 11.675$ M

No. of piles provided in the breasting dolphins = 33 each of 1 m dia. in 6 rows, each row consisting 5 piles + 3 Fender piles of 1.2 m dia.

Assume that each of the piles will carry equal shear. Hence, horizontal load in each pile will be $\dfrac{360}{33}$ = 10.91 T. Taking the moment at the point of contra flexure, we have, induced BM in each pile = 10.91 × 11.675 = 127.32 tm.

2.7.1 Dead Load Reaction

(a) Total dead weight of the pile cap

$$= 18.5 \times 16 \times 1.5 \times 2.5 + \frac{9+16}{2} \times 9.5 \times 1.5 \times 2.5$$

$$= 1110 + 211 = 1321 \text{ T}$$

Add self wt. of fire water monitor tower = 5.0 T, making a total of 1326 T.

(b) Self wt. of Pile

$$= (5.5 - 3.5) \times \frac{\pi}{4} \times 12 \times 2.5 + (3.5 + 26.5) \times \frac{\pi}{4} \times 12 \times 1.5$$

$$= 3.9 + 35.1 = 39.0 \text{ T}$$

Hence, total dead load on a pile would be:

$$= \frac{1321}{33} + 39 = 40 + 39 = 79.0 \text{ T}$$

2.7.2 Design Particulars

Now the pile can be designed for a BM of 127.32 tm and a direct reaction of 79.0 T.

- Design the pile using ultimate load theory.
- Adopt a concrete mix of M_{35}
- Use TOR steel

$$= \frac{P_\mu}{\sigma cu.D^2} = \frac{79 \times 1.85 \times 10^3}{350 \times 100 \times 100} = 0.042$$

$$= \frac{M_\mu}{\sigma cu \times D^3} = \frac{127.32 \times 1.85 \times 1000 \times 100}{350 \times \dfrac{\pi}{4} \times 100 \times 100 \times 100} = 0.086$$

Use a cover of 60 mm $\dfrac{O'}{O} = \dfrac{88}{100} = 0.88$

Hence, from column interaction dia. $\dfrac{r}{\sigma cu} = 0.70$

Hence, At = $350 \times 0.7 \times 10^{-4} \times \dfrac{\pi}{4} \times 100 \times 100 = 191.1 \text{ cm}^2$

Provide 32, TOR

A_t provided = $24 \times 8.04 = 192.96 \text{ cm}^2$ as against 191.1 cm² required.

Hence, OK

I of the pile $= \dfrac{\pi}{64} \times 100^4 + \dfrac{(19 \times 1.5 - 1) \times 192.96 \times 88^2}{8}$

$$= 49,06,250 + 51,36,595$$

$$= 100,42,845 \text{ cm}^4$$

$$z = \frac{100,42,845}{50} = 2,00,857 \text{ cm}^3$$

2.7.3 Induced Stresses

Induced stresses are:

$$\text{Bending} \quad = \quad \frac{127.32 \times 10^5}{200857} = 63.40 \, \text{kg/cm}^2$$

$$\text{Direct} \quad = \quad \frac{79 \times 10^3}{\frac{\pi}{4} \times 100 \times 100} = 10.02 \, \text{kg/cm}^2$$

$$\text{Now,} \quad \frac{\sigma c'}{\sigma c} + \frac{\sigma b''}{\sigma b} = \frac{63.40}{115} + \frac{10.02}{90}$$

$$= \quad 0.55 + 0.11$$

$$= \quad 0.66 < 1 \text{ OK}$$

From the above, it is evident that the first part of the crack theory is satisfied.

The resultant tension being (–) 63.4+10.02 = (–) 53.38 kg/cm², the crack theory is not fully satisfied. However, considering the nature and duration of the impact force and the magnitude of the tension induced, the cracks would remain within the tolerable limits.

If, however, it is desired to have more a rigid structure, raker piles may be introduced in addition to the vertical piles and the pile bend analysed using STAD-III programme.

3

Design of Jetty Fendering System

3.1 INTRODUCTION

During the process of berthing of a tanker alongside of a jetty, it may hit the jetty. In LPG/LNG jetties, as said in Chapter 2, the mooring dolphins and the central unloading platform (CUP) are set back off the berthing face. As such and considering the fact that the angle of approach of the tanker is limited to 3° or at the most to 5°, the chances of tanker hitting to these structures are remote. The remaining jetty structures, viz., breasting dolphins, which are in line with the berthing face, may receive impact of berthing tankers. As such it is very essential to equip the jetty with a suitable fendering system. The aim and objective of providing the fendering system in the berthing face of the breasting dolphins are to absorb the high impact energy induced by the berthing tankers so as to transmit a low reaction force to the jetty structure and also to tanker's hull. In other words, the fendering system serves as a bumper to protect the jetty and tanker's hull from any possible damage.

A tanker approaches a jetty at an angle with a certain approach velocity. It then strikes a breasting dolphin, skids a little distance along the dolphin face and is made parallel to the berthing face. Unless carefully manoeuvred, the tanker may give a violent blow to the first dolphin followed by a similar blow to the next dolphin. In any case, the berthing forces transmitted to the breasting dolphin consist of an impact force normal to and a frictional force parallel to the berthing face.

3.2 MAGNITUDE OF KINETIC ENERGY

When an impact occurs some energy is absorbed by the elastic deformation of the dolphin, some by elastic deformation of the hull structure, some energy is dissipated by rotation of the tanker and the balance energy is absorbed by compressing the fender unit. This kinetic energy can be derived from the following expression:

$$E = (W_D \times C_m) \times C_e \times C_s \times \frac{V^2}{2g}$$

Where,

E = Kinetic energy in tm.

W_D = Displacement tonnage (DT) in tonnes

g = Acceleration due to gravity (9.81 m/sec)

C_m = Mass coefficient

C_e = Eccentricity coefficient

C_s = Softness coefficient

3.3 DISPLACEMENT TONNAGE (DT)

For tankers, the relationship between DT and DWT generally are as follows:

DWT	25,000	50,000	80,000	1,00,000	1,25,000	2,25,000 & above
DT/DWT	1.32	1.26	1.25	1.20	1.17	1.15

3.4 HYDRODYNAMIC MASS COEFFICIENT

When a moving tanker is brought to rest, a body of water moving with the hull is also brought to rest which increases the hydrodynamic weight of the tanker. This additional weight plus the displacement tonnage of the tanker constitute the virtual weight of the tanker which is taken to assess the berthing energy. For calculating the virtual weight, a mass coefficient (C_m) is employed which is defined as under:

$$C_m = 1 + \frac{2D}{B} \qquad \qquad \text{... (1)}$$

Where,

D = Draft of the tanker in m

B = Beam of the tanker in m.

Alternatively, the additional weight may be taken as the weight equal to a cylinder whose diameter is equal to the draft of the tanker and whose length is equal to length of the tanker. Hence:

$$\text{Additional Weight} = \frac{\pi}{4} \times D^2 \times L \times W_0 \qquad \qquad \text{... (2)}$$

Where, W_0 = unit weight (density) of water/m^3 which can be taken as 1.03t/ m^3.

Hence, virtual weight of the tanker becomes:

$$W_v = (W_D \times C_m) \text{ or}$$

$$W_v = (W_D + \frac{\pi}{4} \times D^2 \times L \times W_0), \text{ whichever is greater} \qquad \ldots (3)$$

Where,

W_v = Virtual weight of the tanker in tonnes

W_D = Displacement tonnage of the tanker in tonnes.

3.5 ECCENTRICITY COEFFICIENT (C_e)

A tanker approaches a jetty at an angle and touches it at a point at the end of its parallel middle body either near its bow or stern as shown in Fig.3.1, then rotates and becomes parallel to the berthing face. By this rotational motion, some kinetic energy is lost, apart from the other losses and the balance energy is considered for the design of the fender block. This is determined by employing a factor (C_e) which can be derived from the following expression:

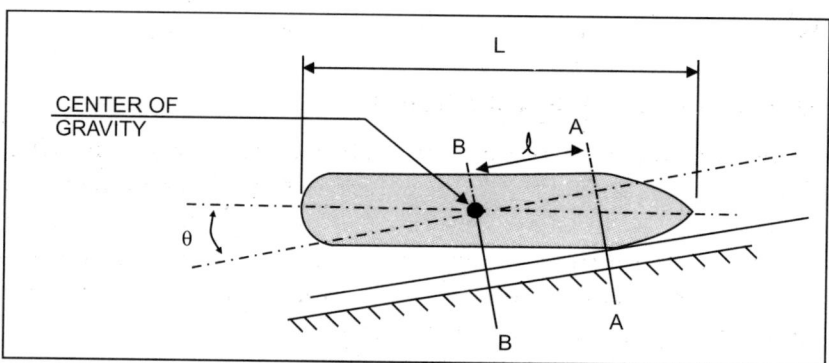

Fig. 3.1: Approach of the tanker on the jetty

$$C_e = \frac{1}{1 + \left(\frac{l}{r}\right)^2}$$

Where,

l = Distance along the water line of the jetty from the centre of gravity of the tanker to the berthing point (m).

r = Longitudinal radius of gyration of the tanker (m).

Radius of gyration can be obtained from the block coefficient of the tanker from (Fig. 3.2).

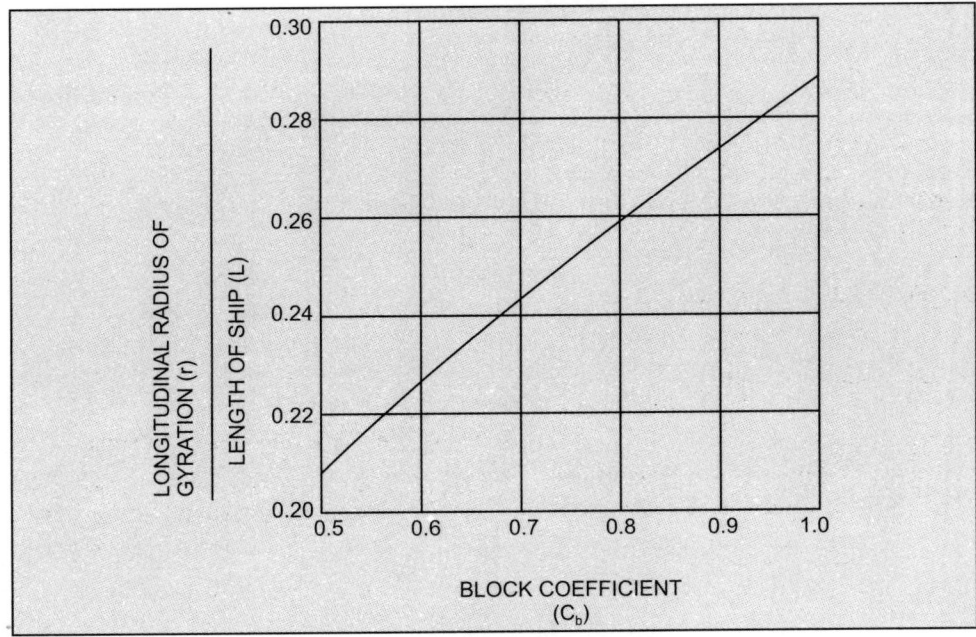

Fig. 3.2: Longitudinal radius of gyration

$$C_b = \frac{W_D}{d.Lpp.B.W_0}$$

Where,

C_b	=	Block coefficient
d	=	Draft (m)
Lpp	=	Length between perpendiculars (m)
B	=	Breadth (m)
W_0	=	Density of water m³ (1.03)

Owing to the large flaring in LPG and LNG tankers, the first contact point of the tanker would be at $1/3 \times$ LOA of the tanker as compared to ¼ point i.e., ¼ LOA in case of a cargo vessel. Consequently, the radius of gyration (r) of a tanker can be approximated to $1/6 \times$ LOA. Hence, corresponding to contact point $1/6 \times$ LOA, C_e works out to about 0.70 for LPG and LNG tankers as against 0.5 for cargo vessels with contact point.

3.6 SOFTNESS COEFFICIENT (C_s)

The softness coefficient (C_s) indicates the relationship between the rigidity of the vessel and that of the fender and a value of 0.9 to 0.95 can be taken. However, some authors approximate it to 1.0.

On the basis of the above analysis and taking of acceleration due to gravity as 9.81 m/sec², we have,

$$E = W_\upsilon \times 0.7 \times 1 \times \frac{V^2}{2 \times 9.81} \qquad \qquad \dots (4)$$

Where,

V = Approach velocity m/sec normal to the jetty.

Each of the fender units must be capable of absorbing this impact energy and the jetty structures in turn must be capable of withstanding the horizontal reaction arising out of the compression of this fender unit by bending of the shaft.

3.7 APPROACH VELOCITY

IS-4651 (Part-III); has specified various approach velocities normal to the berth, which are reproduced below.

Sl. No.	Site Condition	Berthing Condition	Berthing Velocity Normal to Berth in m/s			
			Up to 5,000 DT	Up to 10,000 DT	Up to 1,00,000 DT	More than 1,00,000 DT
(1)	(2)	(3)	(4)	(5)	(6)	(7)
(i)	Strong wind and swells	Difficult	0.75	0.55	0.40	0.20
(ii)	Strong wind and swells	Favourable	0.60	0.45	0.30	0.20
(iii)	Moderate wind and swells	Moderate	0.45	0.45	0.20	0.15
(iv)	Sheltered	Difficult	0.25	0.20	0.15	0.10
(v)	Sheltered	Favourable	0.20	0.15	0.10	0.10

3.8 STATISTICAL METHOD

It is seen that owing to several indeterminate factors, the berthing energy evaluated by the theoretical method given in para 3.2 may be lower than what has actually been measured at some jetty sites which includes the cumulative effect of the berthing velocity, hydrodynamic mass and various, eccentricity factors. Figure 3.3 shows one curve based on measurements of energy at Rotterdam and other two curves are based on BS Code of Practice and the Norwegian Standard for berth structures. The Norwegian Standard also mentions that for harbours exposed to strong winds and currents or with particularly difficult manoeuvring conditions, the impact energy given above shall be increased up to 50%. For structures in the open sea the impact energy shall be increased by 100%. Hence, in order to keep the impact at a low level, an LPG/LNG jetty should be located in a tranquil basin with an easy approach from the turning circle and manoeuvred by tugs with an approach angle not exceeding 3° or at the most 5°.

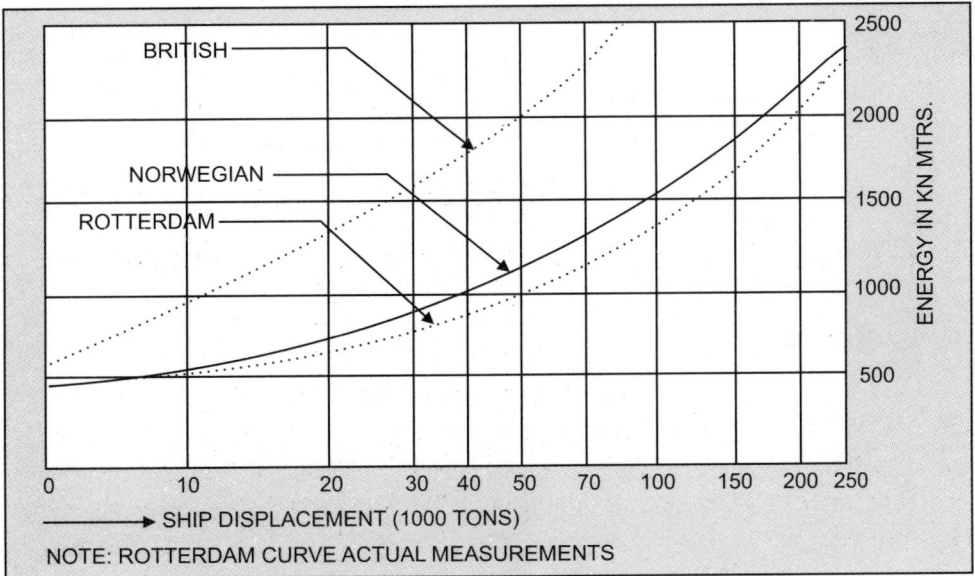

Fig. 3.3: Impact energy during ship berthing to a berth structure

However, depending on the location and exposure of the jetty structure
to wind, wave and current, designer should exercise his own judgement
while determining the berthing energy.

3.9 TYPES OF FENDER

In light of the above analysis, it is evident that an appropriate fendering
system shall have to provided in the breasting dolphins to protect the LPG/
LNG tanker and also the jetty structure against possible damage. Many types
of dock fenders in various rubber grades are being manufactured in India
and abroad. These fenders fall into the following broad groups:

• Cylindrical rubber fenders including side loaded hollow cylindrical fend-
  ers and end loaded hollow cylindrical blocks.
• V, M and arch shaped fenders.
• Cell fenders with or without a protective frontal frame.
• Pneumatic (air/foam filled) fenders.
• Figure 3.4 shows various types of fenders. Cylindrical and pneumatic
  fenders fall in category of soft fenders; Cell fenders are medium fenders;
  V, M and Arch fenders are hard fenders.

3.10 PERFORMANCE CURVES

All the fender manufacturers give in their published brochures rated and
maximum capacities with regard to energy absorption and the reactive force
thereof, corresponding to a range of specified deflections with a tolerance
limit of ± 10%. Once the berthing energy is known, the characteristic curves

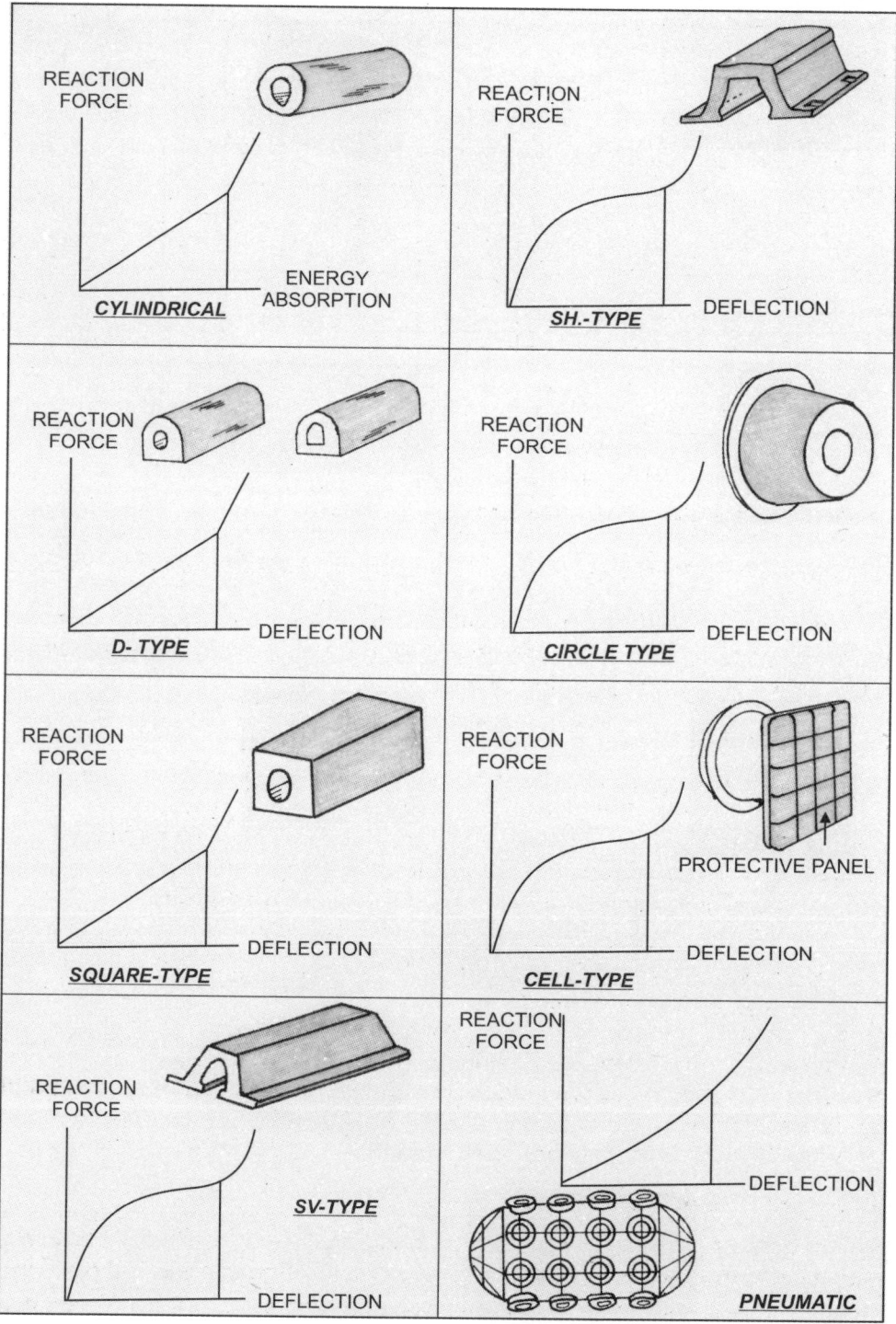

Fig. 3.4: Various types of rubber fenders

can be used to evaluate the reaction forces on the jetty and the jetty structure can be designed accordingly.

3.11 RELATIVE ADVANTAGES / DISADVANTAGES

The relative advantage of the above category of fenders broadly are as follows:

Type	Advantages / Disadvantages
Hollow Cylindrical	**Advantages** (a) It is a soft fender. (b) Economical for small tankers. (c) Easier to install or replace individual units. (d) Can be mounted suitably to give cover to a large vertical range. **Disadvantages** (a) Susceptible to damage by surging motion of tankers. (b) Long length necessary to spread the reactive force. (c) Larger length is required to absorb same energy. (d) Does not prove economical specially for large tankers.
V, M and Arch Fenders (Bucking fenders)	**Advantages** (a) These are hard fenders. (b) Easy to install and replace individual units. (c) Low reaction and high energy absorption over a part of the deflection range. **Disadvantage** (a) Only covers a small tidal range unless upper and lower rows of fenders are provided.
Cell Fenders	**Advantages** (a) These are medium fenders. (b) Easy to install and replace. (c) Simple in design. (d) Large energy absorption capacity. (e) Several units can be coupled to a common fender frame to give additional energy capacity.

Type	Advantages / Disadvantages
	(f) With a frontal frame, transmits a low reaction force to hull structures of the tankers.
	Disadvantages
	(a) Required a frontal frame even for a single unit.
	(b) Requires chains to help to carry both weight and longitudinal forces.
	(c) Unless special care is taken, under water fixtures are liable to corrode specially the embedded bolts in concrete requiring removal and replacement at some stage which are difficult, expensive and may put the jetty out of action for few days.
Floating Pneumatic Fenders	**Advantages**
	(a) These are soft fenders.
	(b) Large energy absorbing capacity.
	(c) Low reactive force.
	(d) Can adjust to curved hull surfaces of the tanker.
	(e) Full tidal length can be covered with a single unit.
	(f) Simple to install and remove.
	Disadvantages
	(a) Large diameter increases the reach of the hard arm.
	(b) Requires a large fender wall or a skirt extending below the dock level and several meters low LLWL.
	(c) Can roll up on to the face of the jetty unless suitably restrained.
	(d) Requires secondary protection against tearing and puncturing by protrusions from tanker's hull and any other sharp iron and steel material.
	(e) Constant designed internal air pressure has to be maintained by a compressor unit.
	(f) Requires spare fenders for replacement involving high inventory cost.
	(g) Fenders may burst and balloon may totally collapse in case there is an abnormal increase in the berthing energy due to an accident or bad berthing which may endanger the safety of a berthing LPG/LNG tanker.

3.12 HULL PRESSURE

The allowable hull bearing pressures for gas tankers should not exceed 15-25 t/m². The German code of practice, however, limits the hull pressures for gas tankers to 20 t/m². The surface pressure of typical fenders are of the following order:

V and M fenders	: 50–140 T/m²
Pneumatic fenders	: 10–25 T/m²
Cell fender with frontal frame	: Varies as per the size of the frontal frame.

3.13 HYSTERESIS

The berthing energy that is absorbed by a rubber fender during compression is partially restored to the mass and partially dissipated inform of heat within the rubber material itself and the later effect is called hysteresis effect and is represented by the area between curves 1 and 2 or 1 and 3 as shown in Fig. 3.5. If the hysteresis effect is very small, fender of this rubber material is a recoiling fender and if a tanker hits a fender of this material it will be thrown out from the fender after it has absorbed the berthing energy. But if the hysteresis effect is large, a tanker will not be thrown out from the fender and will slowly be pushed out from the fender without any sudden kick or jerk. Further, fenders with small hysteresis acting in conjunction with the

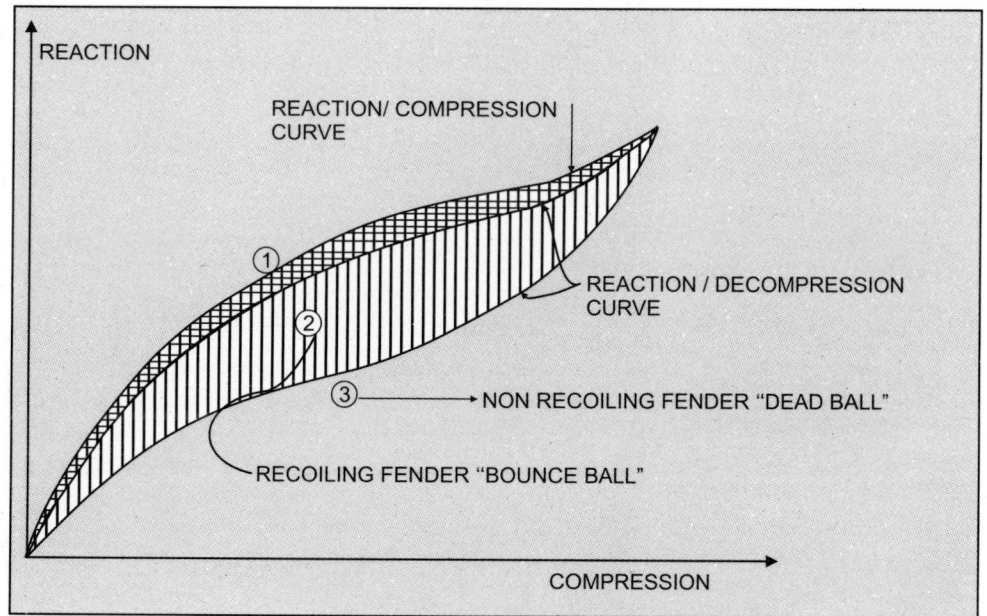

Fig. 3.5: Hysteresis effect

lateral forces due to wind, wave and current acting abeam to the tanker, may cause a resonating condition wherein the tanker may move out of the jetty face and again come back at a particular resonating period. This sway motion to and fro may adversely effect the ship to shore transfer of LPG/LNG and may adversely strain the mooring ropes. Hence fender material with large hysteresis should be used. Cell and Air block fenders have large hysteresis effect and therefore released reactions are slow to push the tanker off the berthing face.

3.14 LIFE OF FENDER

The life of a fender block, is generally around 5–15 years depending on the maintenance. However, fenders are damaged either due to under-dimensioned, bad manoeuvring or too high a berthing velocity, surge and roll motions of the berthed tanker and lack of proper maintenance. Fender fixtures and chains need annual maintenance and replacement regularly. The fender wall or the fender skirt, after a period of 5–10 years suffers extensive damage due to internal corrosion of the reinforcement because of which concrete falls off. Since this happens in under water portions, it becomes very difficult to repair the damaged concrete skirt. The skirt being an integral part of the jetty neither can be replaced nor can be repaired satisfactorily. It is therefore suggested that the fender wall should be of robust design, directly resting on a group of piles and should be of dense concrete of mix nor less than M_{35} with a minimum cover of 60 mm and should be designed using elastic theory with working stress method, stress in steel limited to 1750 kg/cm^2 and in concrete (M_{35}) 90 kg/cm^2 (direct) and 115 kg/cm^2 for bending. All the fender fixing bolts, nuts and washers that are to be embedded into the concrete should be of high grade stainless steel and no part of these bolts shall come in contact with the reinforcement of the fender skirt.

3.15 ANGULAR COMPRESSION

The angular compression of a fender block may occur due to the following factors:

- Too large angle of approach between the tanker and the fender line.
- Curve of the hull of the tankers in plan where tanker makes contact with the fender.
- Flare angle of the hull in section.
- Surge, heave and roll movements of the tanker along side of the fender line.

The fender manufacturer's curves are for uniform deflection of the fenders. But when above factors are considered the fender would undergo an angular compression and will also be subjected to certain amount of torsion. A correction factor need to be applied while evaluating the energy of the fender. Most of the fender manufacturers give rated energy absorption for various angular compressions. The energy absorption decreases under the angular compression and the manufacturers recommend to limit the deflection of the fender to about 40–45° as against 52.5° normally allowed. Following values as given in Table 3.1 may be adopted.

Table 3.1: Angular compression		
Angular compression	**Reactive force**	**Absorbed energy**
15°	0.88	0.69
10°	0.89	0.78
5°	0.93	0.88

For LPG/LNG tankers, the maximum angular compression may be limited to 10°.

3.16 LONGITUDINAL FORCE

In addition to the transverse force due to the elastic compression of the fender, a longitudinal force also occurs simultaneously due to skidding of the tanker after the first impact, when the fender is still in compression state. This force which is transmitted to the jetty structures via the fender is in form of a frictional force, $F = \mu P$ where μ is the coefficient of friction between the ship and the fender and P is the force generated due to the impact. The frictional coefficients are steel to steel 0.25; timber to steel 0.4 to 0.6; rubber to steel 0.6 to 0.7; and steel to anti-friction pads 0.15 to 0.20. However, to take care of this longitudinal force, a force equal to 20% of the transverse force should be taken into account and the jetty structure should be designed accordingly for both the forces acting simultaneously.

3.17 MEANS TO REDUCE BERTHING IMPACT ON FENDERS

To high an approach velocity, too steep an angle of approach and bad berthing manoeuvre coupled with human errors, strong unfavourable winds and currents may give rise to a violent impact which may seriously damage the fenders and the hull may come in contact with concrete face of the dolphin which in turn may cause serious damages to the hull structure. In order to avoid an hazardous accident, tug boats should invariably be used while manoeuvring the tanker. The berthing should be made as gentle as feasible.

Prof. Vasco Costa studied the basic berthing manoeuvres and have given valuable suggestions for safe berthing of vessels alongside of a jetty and a wharf. These studies can also be applied while conducting berthing manoeuvres for the gas tankers.

The berthing manoeuvre should be so conducted that the first contact of the tanker with the fender of a breasting dolphin can be established at the end of its parallel middle body. In case of a fully laden tanker, the centre of gravity is nearer to the bow than to the stern. If the stern is made first to touch the fender without fouling the propeller with the jetty structure, the CG will be at a greater distance and hence a greater moment can be given to the tanker with lesser force. However, in case of a tanker in ballast, CG is nearer to the stern than the bow. In such cases berthing bow first can be preferred. In either of the cases, off centre berthing of a tanker to a pair of dolphins should be avoided as this may give occasion to a violent second impact to the other dolphin.

In light of the above analysis, we may now discuss the berthing of a LPG/LNG tanker along a jetty. Depending on the location of the turning circle, the tanker can be brought with tug assistance, stern in first and made parallel to the berthing face of the jetty so that the bow faces deep water and preferable wind ahead. The port side or the starboard side can be the along side depending on the location of the jetty and deep water ahead. By method of parallel berthing and with the help of tugs, the tanker can be brought to about 40 m off to the jetty face and can be made parallel to the berthing face of the jetty with the help of two tugs and two limit transits. One spring line and two stern lines can then be passed and secured to the mooring dolphins and the tanker's centre is made to coincide with the jetty centre. By heaving the aft ropes as would be required and with the help of two attending tugs, the tanker can be slowly turned and bow can be made to touch the fender block limiting the angle of approach to 3° or at the most 5° and thereafter the tanker is made parallel to the berthing face by again giving a turning moment to the tanker so that it touches all the fender blocks. Thereafter, the bow can be secured by mooring ropes. By this process, berthing would be smooth and gentle thereby avoiding any violent impact to the fender blocks. This otherwise is known as parallel berthing method.

3.18 DESIGN EXAMPLE

Design a suitable fendering system for servicing an LPG tanker of the following characteristics:

Capacity (T)	DWT (T)	LOA (M)	BEAM (M)	DRAFT (M)	LPP (M)
75,000	46,900	229	36	12.1	218

Assumptions
Angle of approach $\quad\quad\quad\quad\quad$ = $\quad$ 5° (max.)
Approach velocity normal to the jetty $\quad$ = $\quad$ 0.15 m/sec.
Assume 1/6 contact point. Then,

$$C_s = 1.0$$
$$C_e = 0.70$$

Sheltering condition $\quad\quad\quad\quad$ = $\quad$ Moderate
Now,

$$C_m = 1 + \frac{+2D}{B}$$
$$= 1 + \frac{2 \times 12.1}{36} = 1.67$$

Hence, $W_v = W_D \times C_m = 75,000 \times 1.67 = 1,25,250$ T
(OR)
Added Wt $\quad = \frac{\pi}{4}.D^2.LOA \times W_0$

$$= \frac{\pi}{4} \times (12.1)^2 \times 229 \times 1.03$$

$$= 27,134 \text{ T}$$

$\therefore W_v = 75,000 + 27,134 = 1,02,134$ T
Adopt max. of the two, i.e., $W_v = 1,25,250$ T

Now, E $\quad = 0.70 \times 1.0 \times \frac{V^2}{2 \times 9.81} \times 1,25,250$ $\quad\quad$ (V = 0.15 m/sec)

$$= 100.54 \text{ TM}$$

It is proposed to equip each of the breasting dolphins with 4 nos. of cell fender units vertically mounted, protected by a frontal frame with anti-friction pads as shown in Fig. 2.5 (Chapter 2).

In providing Shibata – 1000 H – CS1 grade cell fenders in each of the breasting dolphins of the LPG jetty to eater to full impact force of 100.54 TM, we have, impact energy/unit = $\frac{100.54}{4}$ = 25.14 TM as against the actual capacity of the unit as 33.0 tm vide Shibata table.

Also from the same table, reaction force = 75 t/unit

Hence, the total reaction force $75 \times 4 = 300$ T @ 52.5% deflection

If $10°$ angular compression is considered the following reductions should be given:

Reaction force $= 0.89 \times 75 = 66.75$ T

Energy $= 0.78 \times 33 = 25.74$ TM as against 25.14 TM required to be given.

Hence, chosen fender unit is O.K. Each of the breasting dolphins should be designed for a reaction force of 300 t for the transverse force and 20% extra for the longitudinal force, i.e., 60 t, both acting simultaneously.

3.19 SPECIFICATION FOR FENDERS

The specifications for rubber fenders have been presented in Annexure 3.1. The fenders shall confirm to the above specifications and should be tested and certified by a competent authority like IRS, Llods, DNV, etc.

ANNEXURE 3.1

RUBBER FENDERS – PHYSICAL PROPERTIES

Characteristics	Requirements	Testing Standards
A. RUBBER		
I Before Ageing:		
1. Tensile Strength	160 kg/cm^2	ASTM-D-412
	180 kg/cm^2 (Pneumatic Fender)	ASTM-D-412
2. Elongation	400% min.	ASTM-D-412
3. Hardness	Max. 77 deg.	ASTM-D-412
II Before Ageing (70°c 96 Hrs. (Air Ageing):		
1. Tensile strength	Not less than 80% of the original value	ASTM-D-573
2. Elongation	Not less than 80% of the original value	ASTM-D-573
3. Hardness	Original + 8 Deg. max.	ASTM-D-2240
III COMPRESSION SET (70°C×22 Hrs. Heat treatment) 700×96 Hrs. (Pneumatic fender)	Maximum 30%	ASTM: D-395
IV ABRASION RESISTANCE	Maximum 1.5 cc	BS: 903 – A9
V TEAR RESISTANCE	Minimum 70 kg/cm	ASTMD -624

B. VISUAL INSPECTION

Rubber shall be homogeneous in quality and shall be free from foreign materials, bubbles injury, cracks and other harmful defects and block shall confirm to manufacturer's dimensions and weight after physical measurements.

C. LOAD DEFLECTION

Deflection of the fender under various incremental vertical loads up to 60% deflection in a suitable compression testing machine should be recorded and a reaction/deflection curve shall be plotted. Such a curve shall confirm to the original curve supplied by the manufacturer.

Contd...

D. OBLIQUE COMPRESSION

The fender block shall be put to oblique compression and reaction/deflection curve shall be plotted as indicated in Para 'c' above. Following correction factors shall be applied to arrive true capacity of fender. These capacities shall not be less than designed values at the specified rated deflection. If, however the manufacture shall supply load deflection curve for oblique compression, the same can be used to determine the fender capacities at a reduced deflection of 45%.

Angular compression	Reaction force	Absorbed energy
15°	0.88	0.69
10°	0.89	0.78
5°	0.93	0.88

For LPG/LNG tankers the maximum angular compression may be limited to 10°.

Design of
Navigational Channels

4.1 INTRODUCTION

Layout plan of a typical shore line harbour protected by two shore connected breakwaters, various navigational channels such as approach channel (outer channel), entrance channel (inner channel), tuning circle (TC), etc., have been shown in Fig. 4.1. In order to evaluate the dredged depths and widths required to be provided in these channels for safe navigation of tankers, it is necessary first to fix up parameters of the maximum size of the tanker that need to be serviced in the proposed port.

To begin with, a hydrographic survey of the area should be conducted and a sounding chart showing available water depths and bed contours prepared. The other field data, viz., wind, wave, tide and current with their magnitude and direction, at least for one complete year, should be gathered and superimposed on the chart. Marine borings at regular intervals should be taken to know the nature and physical properties of the sub-soil strata. A shallow seismic survey should be conducted to ascertain level and depth of bed rock if any. The navigational channels can then be designed as covered in this chapter and a provisional harbour layout plan prepared considering first the navigational safety, in consultation with the local pilot. The layout plan can then be put to a geometrically similar model test (mathematical or physical) and tranquillity conditions, pattern of siltation and erosion in various zones of the proposed harbour determined and the harbour layout plan adjusted accordingly for a minimum siltation criteria.

4.2 DRAFT REQUIREMENT

The harbour should offer adequate draft commensurate with the largest size tanker that is proposed to be handled. The required draft should be deter-

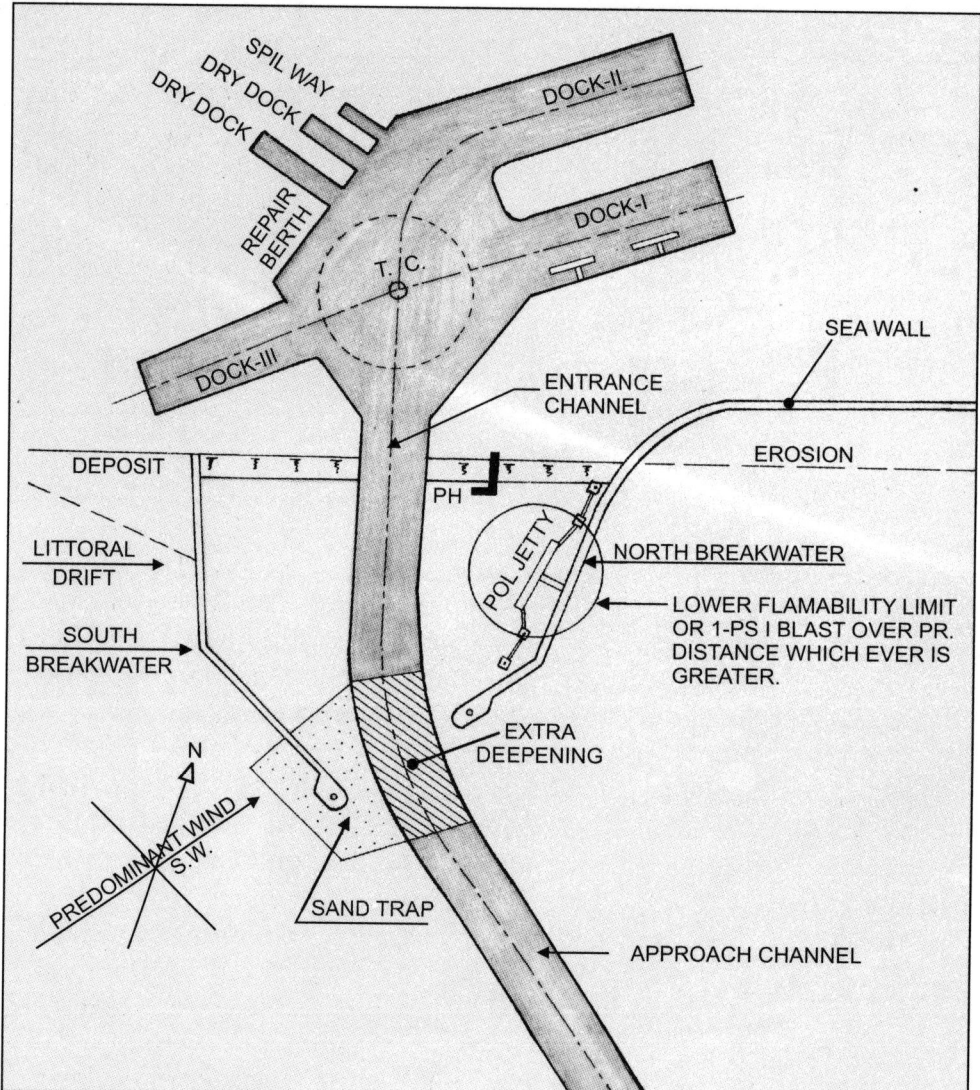

Fig. 4.1: Typical layout of a shore line harbour

mined taking into account of the following factors which are also shown in Fig. 4.2.

- Draft of maximum size fully loaded tanker
- Tidal variations
- Density change
- Squat
- Pitching and rolling

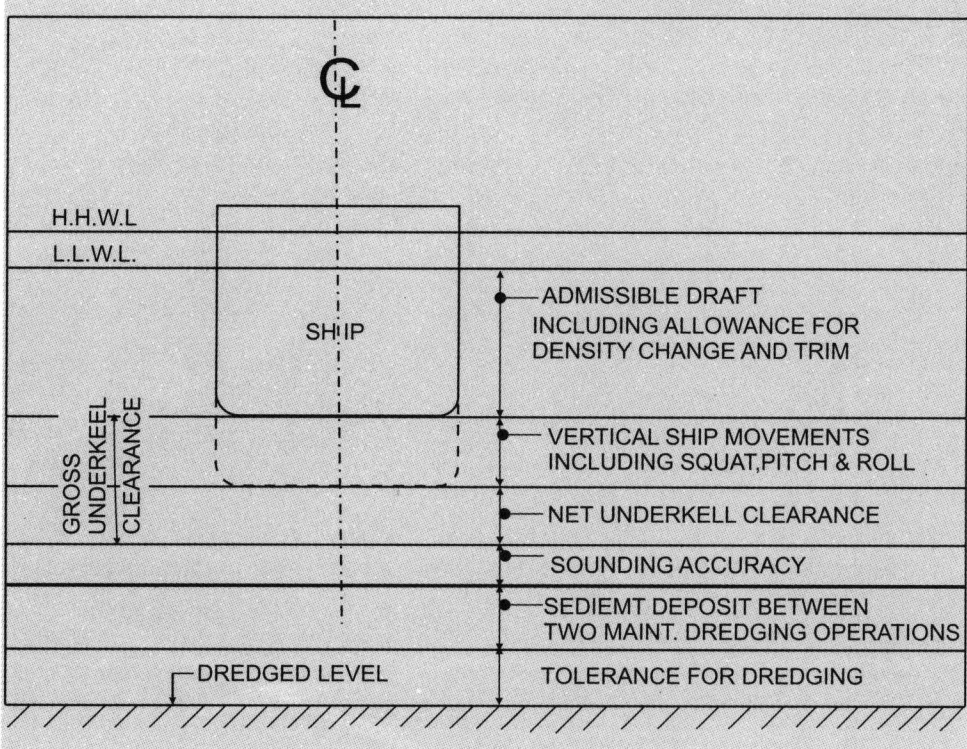

Fig. 4.2: Components of depth (PIANC)

- Trim and
- Empirical factors such as
  - ✦ Sounding accuracy
  - ✦ Allowance towards siltation between two consecutive maintenance dredging operations.
  - ✦ Dredging tolerance, i.e., towards unevenness of the bottom.

4.3 LOADED DRAFT

The maximum size of the tanker that need to be handled has to be fixed after considering the shipping trend that would likely to occur in the near future and also the overall shipping economics. The required harbour can be developed in two stages: first stage could cater to say 60,000 DWT tankers and in the second stage the harbour can be expanded to receive say 1,00,000 DWT tankers. However, it is advisable that from the initial stage itself, the TC should be panned for the ultimate stage of development, although the dredged depths can be increased in stages depending on the requirement that would arise from time to time.

4.4 TIDAL VARIATIONS

The dredged depth should be reckoned from Lowest Low Water Level (LLWL) and that the tanker can be taken inside the harbour in high tide conditions so as to get 1 to 2 m of extra tidal depth which will give greater keel clearance thereby improving propeller efficiency and manoeuvring ability of the tanker. However, while estimating the required draft, this additional draft which is a safety measure, need not be taken into account.

One way to reduce the dredging cost to some extent would be to provide (if the tidal range is high), deeper depths at the tanker jetty and lesser depths in the channels and TC. Fully loaded tankers can be brought into the TC and alongside of the jetty at high water conditions availing extra tidal depths so that the tanker can work alongside the jetty at all stages of tide. After discharging the cargo, the tanker would become light and can sail off at any stages of tide. By this process quantity of capital dredging can be reduced to some extent. However, the drawback in this system is that tanker has to wait at the fairway buoy till high tide condition is obtained. The detention period of the tanker may attract demurrage payments. However, this can be minimised by regulating suitably the expected time of arrival (ETA) of the tanker.

4.5 DENSITY CHANGE

This parameter is mainly applicable for estuarine harbours where fresh water meets the saline water. For sea ports, the effect of density change may not arise.

4.6 SQUAT

When a tanker enters a shallow water region, there is a rapid increase in the height of waves produced by the tanker. Accompanying this increase in the wave height, there is an average decrease in the water surface along the profile of the tanker, relative to the still water level. This surface depression causes the tanker to sink or squat relative to the channel bottom. The factor that affect amount of squat are:

- Clearance between the keel and the bottom
- Trim of the tanker
- Cross-sectional area of the tanker and the channel and whether it is located in a wide or narrow waterway
- Whether the tanker is passing or overtaking another vessel.
- Location of the tanker relative to the centre line of the channel.
- Characteristics of the tanker itself.

Several methods are available for determining the squat specially when the tanker traverses in the centre line of the channel. Most comprehensive method of determining squat is the equation developed by David Taylor Model Basin and Schijf, correlating Froude's number with that of ratio of mid ship cross-section to the channel cross-section and also several other factors. Squat curves that emerged from the experimental studies conducted by Model Basin, have been presented in Fig. 4.3, which can be used to estimate the squat. For example for a tanker speed of 6 knots, squat would be around 1 ft., i.e., 30 cm.

To reduce the squat, keep the sailing vessel as far as possible in the centre line of the channel and give adequate keel clearance.

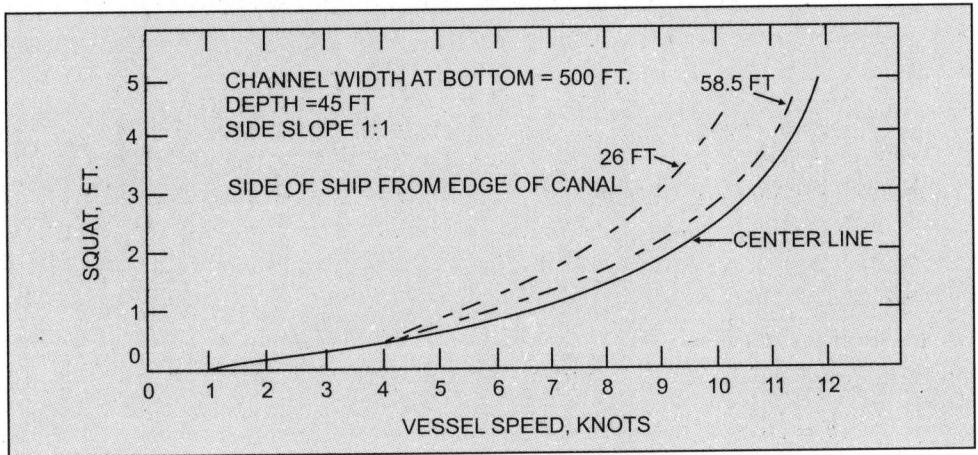

Fig. 4.3: Determination of squat

4.7 TRIM

Invariably, the tanker is not loaded to an even keel and is set down at the stern to obtain some additional depth so as to improve tanker's steering ability. According to Elisiminger, the vessel usually is set down at the stern approximately 3 inches for every 100 ft. (25 mm per 10 m). This means that trim for a 250 m long vessel, would be around 62.5 cm. This empirical rule with regard to the trim can be taken into consideration while assessing the required draft.

4.8 PITCHING AND ROLLING

Additional depth would be required if the tanker is subjected to pitching and rolling, due to the wave action that would be experienced when the tanker is negotiating a shallow water region specially around the entrance zone. Such vertical movements under the influence of wave and swell depend on the wave height, direction and ratio of wave length to the tanker. In long wave periods, specially when the wave period is around 14 secs. and above and in head seas, pronounced heave motion and in beam seas, greater roll motion will be experienced.

As on date, no precise mathematical expression is available to estimate the additional depth that needs to be provided to counteract rolling and pitching. However normal practice is to provide an allowance of half of the wave height that the tanker would be subjected to as suggested by Quinn. The magnitude of roll and pitch motions is also expressed in terms of degrees. If the amount of pitching is known for the design tanker under the design conditions, the amplitude of pitching can be determined knowing the length of the tanker. As regards to roll motion, a 5° amplitude of roll is not uncommon at harbour entrances. This may cause an increase of draft by about 1.22 m for a tanker of 30.48 m beam.

The allowance towards roll and pitch motions in outer channel facing open sea conditions will be greater as compared to the inner channel as it is adequately protected. Newland suggests that due to pitching and rolling of ships, a keel clearance of 10′ to 12′ (3.05 m to 3.66 m) would be desirable for large vessels in open sea waters prior to reaching the lee of a breakwater or a protected channel.

In light of above, allowance towards rolling and pitching both in the inner channel and outer channel should be decided taking account of the wave climate (prevailing round the year) and also the capital and maintenance dredging that would be involved, in consultation with the Pilot of the Port. It may be noted that shipping tugs will not be able to reduce the roll and pitch motions. If a compromise cannot be stuck down with the Pilot, who will navigate the vessel, it would be preferable to take a shutdown of the navigation during the periods when wave height in the outer channel is greater than say 3 m and wave period greater than say 10 secs.

4.9 EMPIRICAL FACTOR

Depth required due to the empirical factors should be fixed taking account of the other factors such as sounding accuracy, allowance towards sediment deposits between consecutive maintenance dredgings and the dredging tolerance. This factor basically, is to reduce chances of the vessel touching the bottom. An allowance of usually 0.6 m to 1.2 m may be given for sandy bottom and lesser speeds and higher value for rocky bottoms and fast speeds. If the port is subjected to littoral drift, it may be advisable to provide an additional depth of 1.5 to 3 m to account for the siltation that may occur due to the drift. Since the littoral drift is at its maximum between the entrance and the breaker zones, this additional allowance may be kept confined to about 300 m inward and 300 m outward of the breakwater tip.

4.10 RECOMMENDATIONS OF PIANC

The recommendations of PIANC, committee for the reception of large vessels, on draft are:

Location	% of Draft of vessel
Open sea	20%
Exposed channel	15%
Sheltered channel	10%

4.11 WIDTH OF NAVIGATIONAL CHANNELS

The width of navigational channels is governed mainly by the steering characteristics of the vessel and its response to the command of the Pilot on board the vessel when it is subjected to the external disturbing forces such as hydrodynamic effects of the bank suction, cross currents, winds, waves and other traffic and also the speed of the vessel. In general steering quality of large vessels in fully laden condition is generally found to be poor specially at slow speeds and with low underkeel clearance. When subjected to strong current, wind and wave induced forces laterally, course deviation in such vessels will be significant which cannot be entirely corrected by rudder action. Such vessels should therefore be manoeuvred with assistance of tugs even if appropriate dredged depths and channel widths are available.

The width of the outer and inner channels should be measured at the bottom and shall comprise of the following lanes (Fig. 4.4). The widths are generally measured in terms of beam (B) of the largest vessel proposed to be serviced:

— Manoeuvring lane (single lane): should be 180 to 200% of the beam for straight channels.
— Bank clearance lane: 75 to 100% of the beam.
— Passage clearance: 100% of the beam for two lane channels.

Thus for a single lane straight channel, bottom width of the channel may be kept between 3.3 and 5 and for two lane channels, 5.1 to 8 x beam of the vessel as shown in Fig. 4.4. Generally speaking for a one-way sailing, unassisted by tugs, a channel width of 5 x beam of the vessel can be adopted which can take care of a 5° yaw due to quartering seas. If higher order of yaw say 10° or more is expected to be experienced, specially in the outer channel, width may be increased to 7B to 8B and tug assistance should invariably be taken, fastening the tugs outside of the breakwater.

The width of the navigational channels can also be fixed in terms of L (LOA) of the vessels. Since length to beam ratio of all the vessels fall within

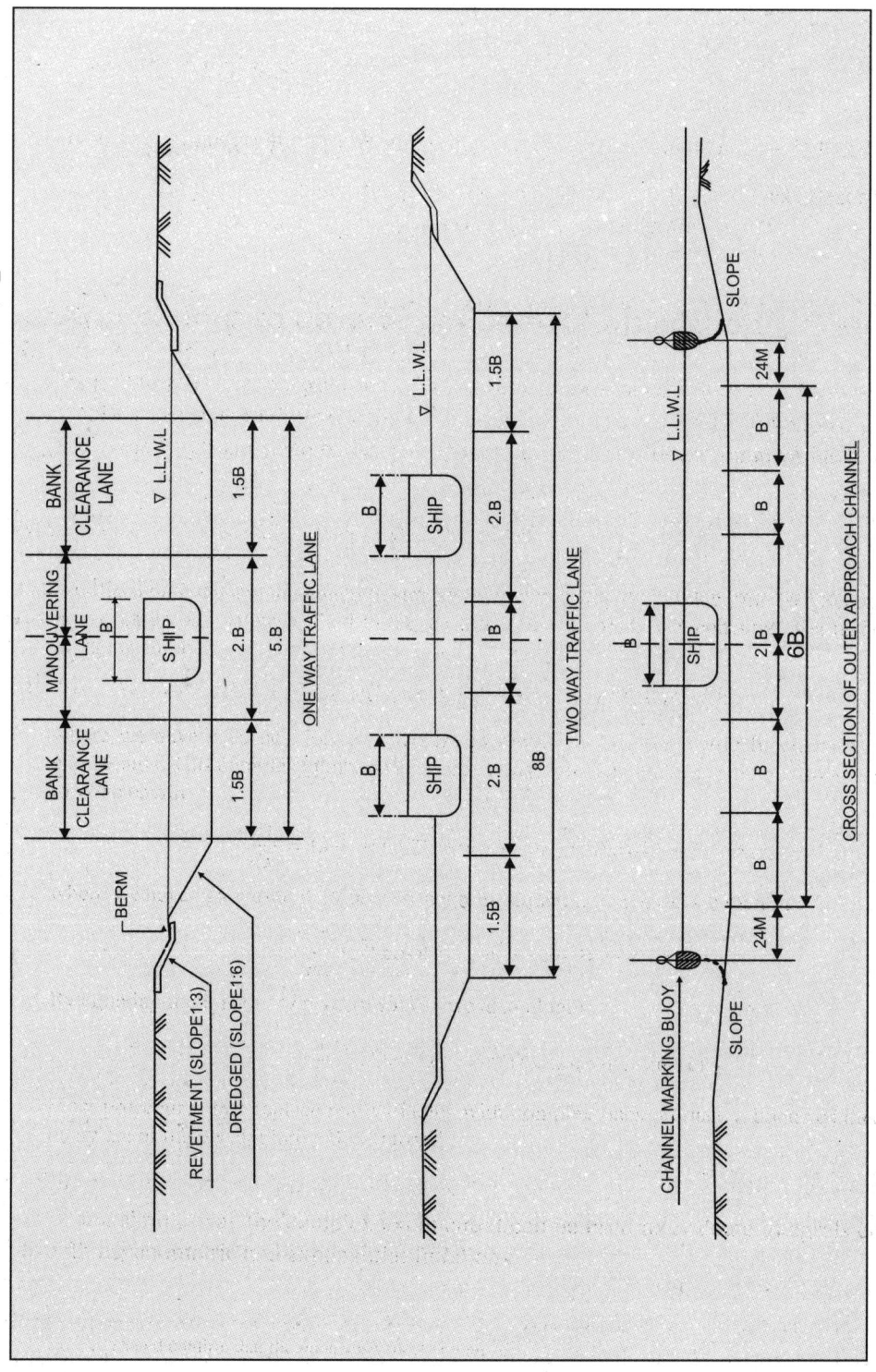

Fig. 4.4: Width of navigational channels

the range of 5 to 7, the width of the navigational channel can be fixed in terms of L, which is equivalent of 5 to 7 beam.

If the outer channel is buoyed, an additional width of about 24 to 40 m should be taken towards displacement of floating buoys that mark the channel in the direction of wind, current or waves, depending on the advice of the Pilot.

4.12 ALIGNMENT OF CHANNELS

Unless dictated by compelling factors such as direction of currents, wind, waves, storms, topography of sea beds, littoral drift and shoaling, etc., the approach channel should be aligned straight and made perpendicular to the bed contours so that capital dredging and the subsequent maintenance dredging involved would be minimum but safety of the navigation should be ensured. The channel should preferably be located in areas of maximum depth to reduce the cost of dredging.

Most of the pilots prefer to handle a vessel in beam waves as the predominant motions that the vessel may experience would be roll and sway. The sway can be corrected by two tugs, one at the aft and the other at the bow on the starboard side if the disturbing forces are from the port side. However, there is no effective means to controlling the roll motion except to navigate the vessel in low amplitude waves limiting the roll motion to 3°. If the vessel is subjected to quartering seas, wind and current, two additional motions will generate, viz., surge and yaw in addition to the sway and roll motions, of which yaw motion is considered to be very dangerous. For unassisted sailings, tolerable yaw motion at the entrance should be limited to 5° and the same in the outer channel to 10°, so that with these limiting yaw motions the vessel would remain within the channel geometry (Fig. 4.5). However, tug assistance should invariably be taken when sailing under strong external forces.

As a general rule, in strong wind conditions, a fully loaded vessel should face wind ahead, which would act some sort of a brake. In a tail wind, the speed will add to the speed of the vessel, which is not desirable as this would require a greater stopping distance.

It is a common practice to make the arrival manoeuvre especially when she is in fully laden condition, more difficult as compared to the departure manoeuvre. This has the advantage that when the vessel is in the light condition, she can leave the port more easily at any time of the day with minimum tug assistance.

The dredgeability of the soil in the channel is another factor which would require careful evaluation after conducting adequate sub-soil exploration.

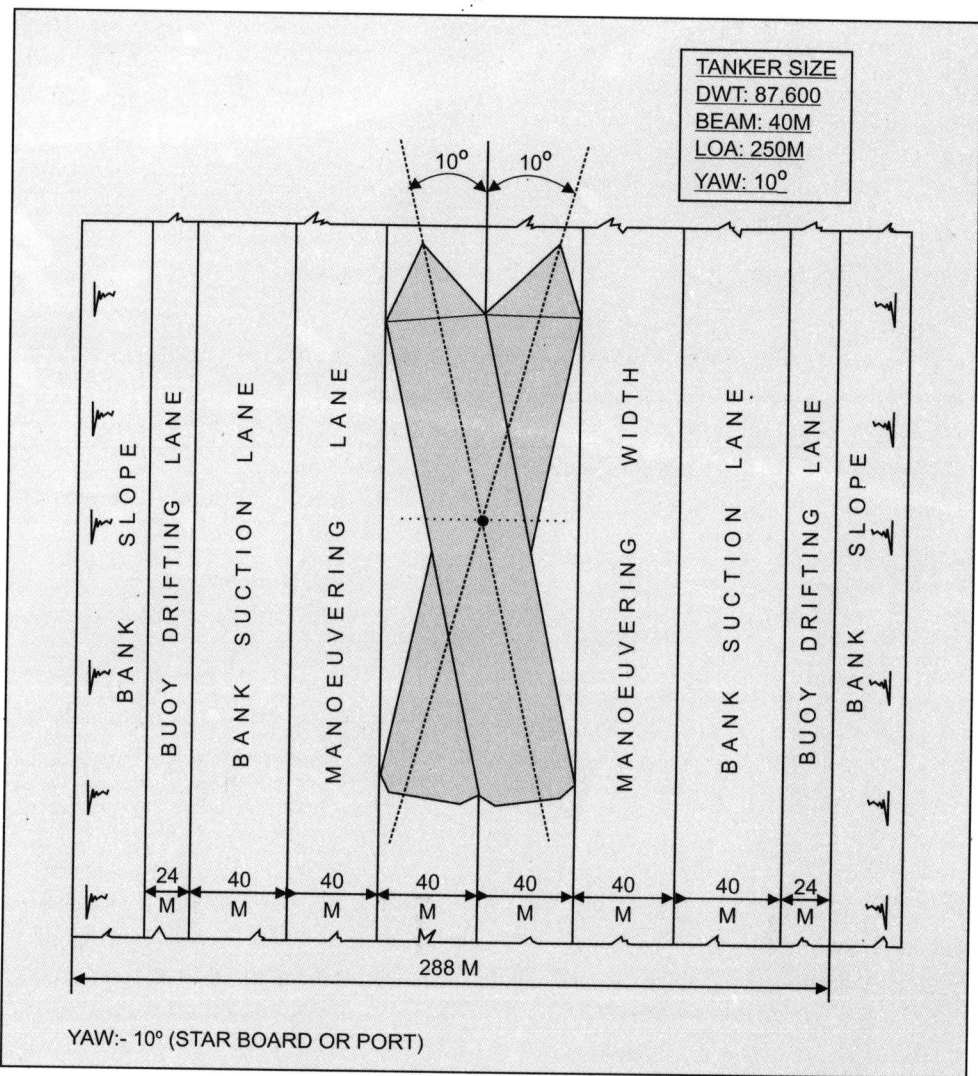

Fig. 4.5: Longitudinal plan of a typical outer channel

Identification of a suitable dumping ground for disposal of the dredged soil from where the dumped soil cannot flow back to the navigational channels, is yet another factor. This is invariably done with the help of model tests (physical or mathematical).

4.13 BENDS IN CHANNELS

In unavoidable circumstances, a bend may be introduced in the channel. The width of the manoeuvring lane in a bend is correlated with the controllability of the vessels and the deflection angle. The deflection, angle is generally limited to 30° although channels with greater angle of deflection of the order of 60°–75° has been adopted. As per 1S-4651 (P-V) the radius of curvature for vessels proceedings without tug assistance should not be less than 3 X L for central angle of the turn up to 25°, 5 X L for turns beyond 25° and 10 L for turns beyond 35° where L is the length of the largest vessel. Where it is not possible to provide above radii, the channel should have to be suitably widened to allow for the swing of the vessel and to provide increased manoeuvring width. As a guide, it desirable to provide following radius of curvature for these bends where possible:

R_{min} = 1200 m for vessels less than 150 m LOA

= 2000 m for 150 m LOA

= 2000 to 3000 m for vessels larger than 150 m but less than 200 m LOA.

A minimum sight distance of 1.5 km should be ensured when a vessel is negotiating a channel bend.

The extra width in the manoeuvring lane that should be provided in the bend has not clearly been established. However, according to an experiment for a 3800 m radius bend, the width of the manoeuvring lane is given as under:

Controllability of vessel	Manoeuvring lane	
	Deflection angle 26°	Deflection angle 40°
Very good	3.25 B	3.85 B
Good	3.70 B	4.40 B
Poor	4.15 B	4.90 B

B = Beam of the largest vessel expected to pass through the channel.

The increased manoeuvring width at a bend can be obtained by employing the following methods:

• Cut off method

• Parallel banks method

• Non-parallel banks method

As recommended in IS-4651 (P-V), the widening should be done by parallel bank method. The slope of the transition should be at least 1 in 20. Two channel marking buoys should be provided at the apex region of the curve as a matter of safety. Many navigators however prefer a series of short tangents connected by short curves which also greatly facilitates while transferring layout plan of the submerged channel on the ground.

4.14 HARBOUR ENTRANCE WIDTH

The harbour entrance width should be sufficient to ensure safe navigation of vessels and at the same time narrow enough to restrict entry of wave energy so as to obtain a tranquil basin in the outer harbour. The navigational requirements for the entrance are related to the size of design vessel, the density of traffic, no. of entrances, the water depth, height, direction and frequency of winds, waves, currents and the littoral drift.

Minikin suggests providing a width equal to the length of the largest vessel expected to enter the harbour. Quinn suggests the following entrance widths:

Small harbour	_	90 m
Medium harbours	_	120 to 150 m
Large harbours	_	150 to 245 m

If the harbour entrance is subjected to a cross current of 1.5 knots, prevailing wave height 1.0 m and the vessel is negotiating the entrance at a controlled speed of not more than 3 knots, a width of 5 X Beam of design vessel can be provided considering an allowable yaw of 5°. If the cross current is 2 knots, wave height is 1.5 m and the vessel speed of 3½ knots, a width of 7 B can be provided considering an allowable yaw of 10°. These widths shall be measured for the maximum draft of the largest vessel at the bed level.

It is not advisable to increase the entry speed of the tanker more than 3 knots while negotiating the entrance specially when the stopping distance is barely sufficient and when there is a shallow patch or a rock bund ahead of TC. If the cross current at the entrance is greater than 1.5 knots, it implies that the channel has silted up due to the littoral drift and that shoals have formed on either sides of the entrance zone. This generally happens immediately after a cyclone. Sounding of this area should be taken up and the channel siltation and the shoals, removed by contract dredging. Failure to ensure this and if the tanker is brought at a higher speed than what has been specified above, in an extreme condition, may lead to grounding of the tanker at the entrance.

Whenever possible the entrance should be located on lee ward side of the harbour. If however the entrance must be located on the wind ward end of the harbour, adequate overlap of the breakwater should be provided so that the vessel should have passed through the restricted entrance and would be free to turn.

The wave orthogonal after refraction should not converge anywhere at the entrance region, thereby avoiding any concentration of energy at the entrance. In fact these orthogonals should diverge at the entrance zone so that calm water would prevail. The sailing direction should therefore be fixed avoiding areas of energy concentration.

4.15 TURNING CIRCLE

A vessel from the entrance channel is admitted into a turning circle (TC) and is then turned with a series of forward or stern movements before berthing alongside of a berth or a jetty. When leaving the berth, the vessel is brought into the TC, turned and is aligned with the centre line of the entrance channel and then is allowed to sail off. The area of the TC would depend on whether the vessel is aided by tugs or not. The generally accepted minimum diameter of the TC is as follows:

- All single screw vessels with tug assistance: 2 X LOA
- All twin screw vessels with tug assistance: 1.7 X LOA

  } Provided full depths are available in the TC at all times

- With no tug assistance, single screw vessels turning by free display of the propeller and rudder: 4 X LOA

The diameter of TC should be measured at the bottom for the maximum draft of the largest vessel.

4.16 STOPPING DISTANCE

From the approach channel, the vessel shall start reducing the speed so that it can be brought to a complete halt within the TC so that it does not overshoot the TC and get grounded in the shallow area ahead. Thereafter, it is turned and is brought alongside a jetty. The distance required to perform this operation is known as the stopping distance of a vessel and will depend on factors such as speed, the displacement/horsepower ratio, reverse power, shape of hull, etc. As a rough guide the following stopping distances may be provided (IS-4651-PV):

- Vessels in loaded conditions: 6 to 8 x LOA

However, when the vessel is assisted by adequate tugs power, a stopping distance of 6 x LOA can be adopted in fairly well sheltered turning circles.

The stopping distance should be measured from the beginning of the protective work to the centre of TC.

4.17 MODEL TEST

The layout plan of the port channels that would be obtained by utilising parameters indicated in this chapter shall be put to a geometrically similar model test, physical or mathematical and tranquillity conditions should be studied. The navigational channel and the gap between the breakwaters can then be optimised till the desired order of tranquillity is obtained. A number of computer manoeuvring simulation models have been developed which can also be utilised to study entry and turning manoeuvres and further optimisation in the harbour layout can be made ensuring safe navigation.

4.18 LITTORAL DRIFT

Economic survival of coastline harbours that are subjected to littoral drift depend on how effectively accretion and erosion arising out of the drift are tackled. In brief this drift is manifested due to oblique approach of waves to the shoreline which generates a long shore current propelling sand particles along the coastline. Due to the construction of the breakwater in the up coast, continuity of the sand movement along the coast is interrupted, causing simultaneous up-coast accretion and followed by down-coast erosion. Eventually, if appropriate remedial measures are not taken, the sand accretion will bypass the navigational channel causing siltation especially at the entrance. The navigational channel would then rapidly loose its depths bringing down a major hindrance to the navigation. In one high intensity cyclone, 6 to 8 m depth of siltation is reported to have occurred in the entrance channel. The north coast side simultaneously under goes rapid erosion endangering shore based properties and infrastructures. Hence, it is of paramount importance to adopt remedial measures right from the inception of the port so as to tackle both siltation and erosion effectively.

The following methods are available to tackle the up-coast accretion and the consequential siltation in the navigational channels and also erosion of the down-coast:

- Installations of a mechanical shore based dredging unit in the up-coast with sand-by-passing system and nourishing the down-coast by dumping the dredged spoil so obtained.
- Construction of a sand trap at the tip of up-coast breakwater, periodical removal of the accumulated silt by a cutter suction/trailing suction dredger and either replenish the down-coast shore or dispose the dredged spoil in to a dumping ground from where the spoil cannot flow back into the channel.
- Construction of a dependable stone revetment work or a sea wall along the down-coast to arrest erosion of the coastline even if beach nourishment is done.

Although, the shore based dredger system has been installed in several east coast ports of India, yet their performance is not found to be satisfactory. Hence, littoral drift may be handled by later two systems, viz., by providing a sand trap at the tip of the up-coast breakwater, removal of the accumulated material by contract dredging and also providing simultaneously a seawall along the down-coast to arrest coastal erosion.

As said earlier, geometrical similar model test (physical or mathematical) should be conducted and pattern of siltation due to the littoral drift at the entrance and in various zones of the harbour basin determined and accordingly, capital dredging and maintenance dredging are regulated.

4.19 AN EXAMPLE FOR CALCULATION OF DRAFT

Design the navigational channels of a shoreline harbour for handling 46,900 DWT LPG tankers given the following data:

Particulars of the design tanker:

DWT = 46,900; Draft Max. = 12.10 m; LOA = 229 m

Significant wave height (Hs) = 3 m; Wave period = 10 sec.

Cross current at the entrance zone = 1.5 knots,

during SW monsoon; Littoral drift = 4 lakh cum.

Sea bed: sandy; LLWL = 0.40 m

Calculation of draft required				
Sr. No.	Items	TC	Inner channel	Entrance channel	Outer channel
		M	M	M	M
A. AS PER CONVENTIONAL METHOD					
1	Draft of the tanker	12.10	12.10	12.10	12.10
2.	Density change	--	--	--	--
3.	Squat	0.30	0.30	0.30	0.30
4.	Trim	0.30	0.30	0.30	0.30
5.	Roll and pitch	0.30	0.30	0.50	0.75
6.	Net UKC	0.50	0.50	0.60	0.60
7.	Empirical factors	0.50	0.60	0.60	0.60
	Total	**14.00**	**14.10**	**14.40**	**14.65**
	LLWL + 0.40	− 0.40	− 0.40	− 0.40	− 0.40
	Dredged levels	− 13.60	− 13.70	− 14.00	− 14.25
	or say	− 13.50	− 13.75	− 14.00	− 14.50

Contd...

Calculation of draft required				
Sr. No.	Items	TC	Inner channel	Entrance channel	Outer channel
		M	M	M	M
B. AS PER PIANC METHOD					
	As per PIANC	1.1×12.1 = 13.31	1.15 × 12.1 = 13.92	1.15 × 12.1 = 13.92	1.2 × 12.1 = 14.52
	LLWL + 0.40	−13.31 + 0.4 = −12.91	−13.92 + 0.4 = −13.52	−13.92 + 0.4 = −13.52	−14.52 + 0.4 = −14.12
	or say	− 13.00	− 13.50	− 13.50	− 14.25

The above calculations reveal that draft calculated using the conventional methods works out slightly higher, as compared to PIANC method. But as the former method is more rational, the same may be adopted.

Tanker Moorings

5.1 INTRODUCTION

LPG and LNG tankers are hazardous in nature. Hence, while decanting these hydrocarbons from a tanker alongside of a jetty, the tanker should be secured firmly to the jetty, whatever may be the severity of the environmental forces that would be existing otherwise the cargo transferring equipment, viz., the unloading arm or the hoses will get adversely strained. The general mooring pattern, the environmental forces and the precautions that should be taken while working alongside of the jetty should be worked out and notified to the pilot of the tanker. Similarly, the written approval of the DC/HM of the port should invariable be taken while choosing the location of the jetty site in the port area and also while designing the mooring system.

5.2 GENERAL MOORING SYSTEM

The mooring system would comprise of aft and forward moorings in which two to three stern/headlines, one or two breast lines and one or two back springs are generally provided depending on the magnitude of forces acting on the tanker. A properly moored tanker is an essential requirement for safe ship to shore transfer of the liquid cargo or vice versa.

The mooring system required to moor a POL or crude tanker alongside of a jetty is identical to the mooring system for an LPG/LNG tanker except that in place of bollards remote operated automatic quick release hooks (QRH) are provided in the dolphins so that in event of a fire on board the tanker or on the jetty, the tanker can be released off the hooks and taken on to the roads quickly.

5.3 CONVENTIONAL PATTERN OF MOORING

A general pattern of mooring by conventional method is shown in Fig. 5.1. The following criteria are to be followed:

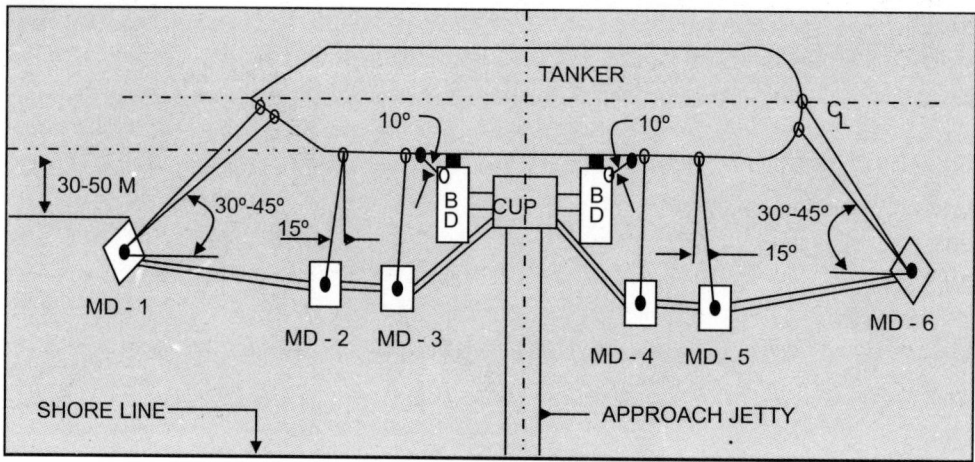

Fig. 5.1: Conventional pattern of mooring

(a) The mooring lines should be arranged as symmetrical as possible about the transverse centre line of the tanker. A symmetrical arrangement is more likely to ensure good load distribution in the mooring lines as compared to an asymmetrical arrangement.

(b) Mooring lines of same size and same material should be used for all the lines in service, i.e., breast lines, spring lines, bow and stern lines.

(c) Length of mooring lines should remain within a range of 35 to 50 m.

(d) Head and stern line angle should be 30°–45° in plan and about 30° in the vertical.

(e) Breast line should be aligned as perpendicular as possible to the longitudinal centre line of the tanker but the breast line horizontal in the plan should preferably be around 15° or less in order to apply maximum restraint to prevent the tanker being moved broad side from the jetty.

(f) Spring lines should be aligned as parallel as possible to the longitudinal centre line of the tanker, angle not to exceed about 10°.

5.4 MOORING BY BREAST LINE

The conventional mooring system as discussed in the preceding para generally comprises of a set of head and stern lines forward and aft breast and spring lines. This type of moorings occupy a large water front which may be difficult to obtain in an enclosed harbour. Secondly, bow and stern lines are normally not very effective in restraining a tanker alongside of a jetty due to their long lengths and poor orientation. According to another system of mooring, stern and head lines can be dispensed with and the tanker can effectively be tied with the jetty by breast and spring lines if the mooring points are properly located and if mooring points allow several hawsers. Further, according to a study conducted by the Centre for Advanced Maritime Studies, Edinburg, UK, restraining tanker with breast and spring lines within tankers own length as shown in Fig. 5.2 would be more effective as compared to the conventional system of mooring with long head and stern lines provided that the jetty is aligned parallel to the direction of predominant forces. However, for easy manoeuvring the tanker, for the purpose of berthing and de-berthing alongside of the jetty, conventional mooring system is preferred even today by some pilots. But, if the port is subjected to high tidal variation or a tidal bore, a long head line need to be provided to reduce the surge and sway action on the tanker.

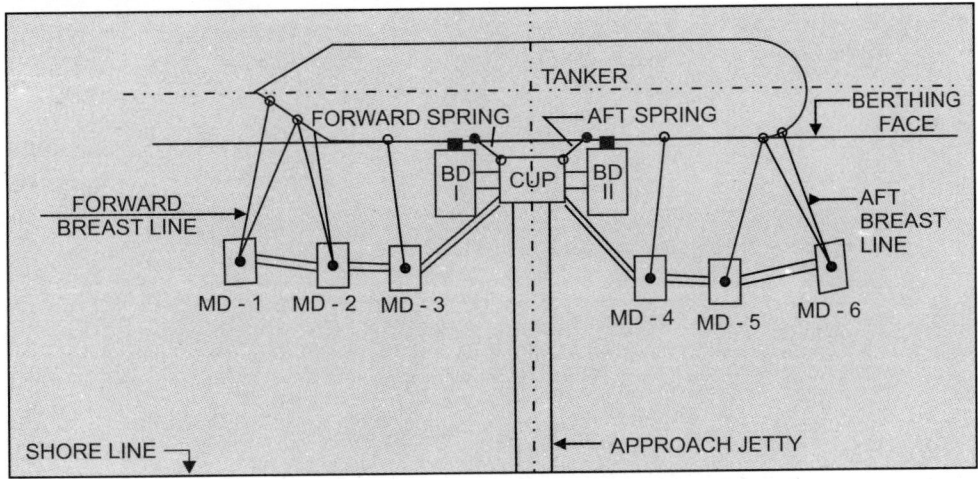

Fig. 5.2: Mooring by breastline

5.5 MOORING FORCES

The tanker's mooring system should be designed to resist all the forces arising out of predominantly from wind, current and wave set up forces. Failure of a mooring system would result undue strain on the hard arm or the hose which may rapture and would cause accidental release of the hydrocarbons with devastating consequences. Hence, a tanker must be moored properly to resist the induced forces, whatever may be their severity.

5.6 WIND AND CURRENT FORCE

The magnitude of the wind force acting on a tanker depends on the velocity, direction and the area exposed to the wind. Wind pressure varies as the square of the wind velocity and linearly with the area exposed of the tanker. When she is on ballast or in light condition, the area exposed to the wind would be greatest and consequently the wind force on the tanker would be highest. When she is fully laden, her exposed area to wind becomes least and accordingly she would experience minimum force. Figure 5.3 shows the various directions of wind and also variation of wind force on the tanker.

It would be evident from Fig. 5.3 that the beam winds will exert maximum pressure on the tanker. The quartering winds although will exert lesser pressure yet the tanker will simultaneously be subjected to a longitudinal force and a transverse force. Tail winds will exert least force. However, the wind acting from the land side would cause considerable amount of sway motion, whereas the quartering wind from the land side would cause both sway and surge motions.

Current force considerations are similar to those that of the wind force. The magnitude of the current force primarily depends on the velocity of the current, the hull area exposed to the current and under keel clearance of the tanker. Maximum force will be induced when the tanker is in a fully loaded condition and the current is acting directly on the beam. Minimum force would be exerted if the tanker is light and its bow is heading into the current.

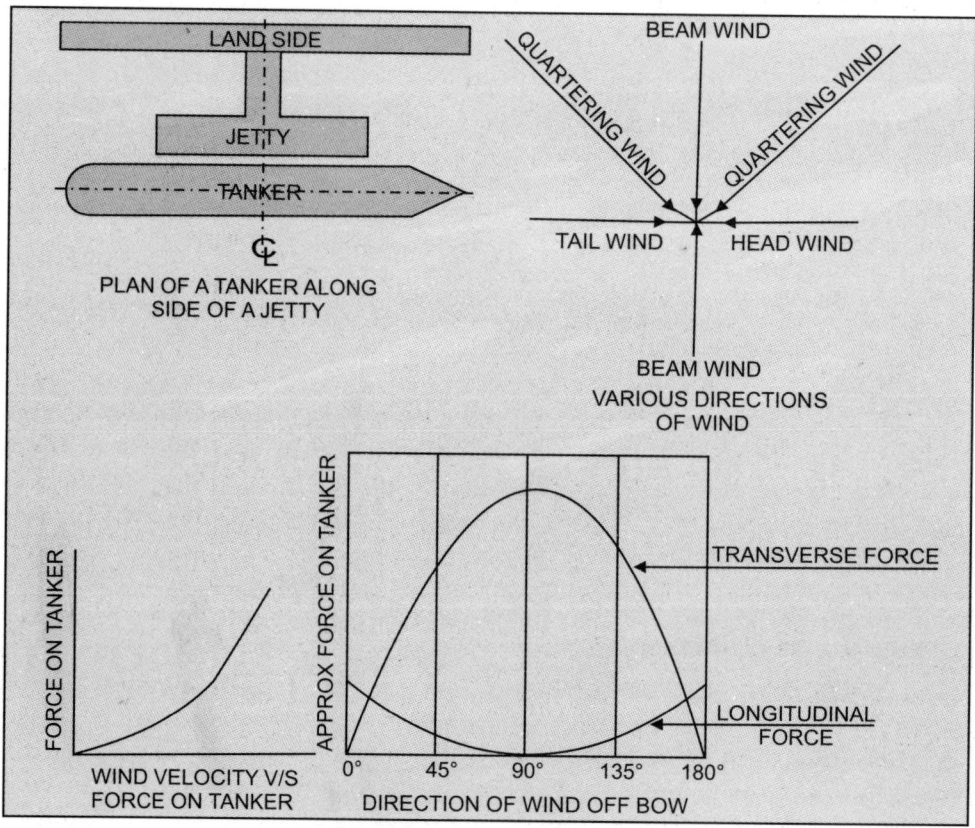

Fig. 5.3: Wind force on tankers

The depth of water under the keel of the tanker can generally affect the current force. As the clearance under the keel decreases, the force due to the current increases. In fact the magnitude of the current force can be three times as great on the vessels having a small under keel clearance than for vessels in deep water.

5.7 WAVE FORCES

Generally speaking, when an LPG and LNG jetty is situated inside a protected basin, the wave disturbance are of low order and therefore wave force on the tanker is not appreciable. But when a jetty is located in more an exposed locality with poor tranquillity conditions, wave forces become significantly high. The wave direction (angle of attack) and the wave frequency (period) are the two factors which generally influence the effect of waves on a moored tanker. There would be an increase in all six degrees of freedom as the wave direction changes head on to the beam. This change would be more significant when the wave direction changes from 45° to 90° even if the tranquillity conditions at the jetty are visibly good. However, a moored tanker is disturbed not by a single wave but by a wave train comprising of several short period waves.

Long period waves (T > 20 secs) may be present at the jetty site, especially when it is built in exposed localities such as an outer harbour. These waves known as seiches are of low height with a long wave period, as high as 60 to 120 sec. and are extremely difficult to detect. If the natural period of oscillation of the tanker at her moorings become the same as that of the periods of the seiches, the moored tanker would move in resonance by which process large mooring forces would be set up in hawsers. Actual measurements of the line loads for moored VLCC'S show that the harbour seiches can cause tanker mooring loads to increase by 15 to 20 t. Hence, such exposed locations would not be suitable for LPG/LNG Jetties and should be avoided.

5.8 TIDAL FORCES

The tidal rise and fall in water level may cause a change of elevation of a tanker with respect to a jetty top level. This will happen when the tanker is loading or discharging. The slacking of or heaving in of the mooring lines should be controlled. Without line tending, increased mooring forces due to tidal raise can be quite severe at some terminals.

5.9 ORIENTATION OF MOORING ROPES

Having discussed the various forces due to wind, wave and tide that can act on an LPG/LNG tanker working alongside of a jetty, we would now discuss

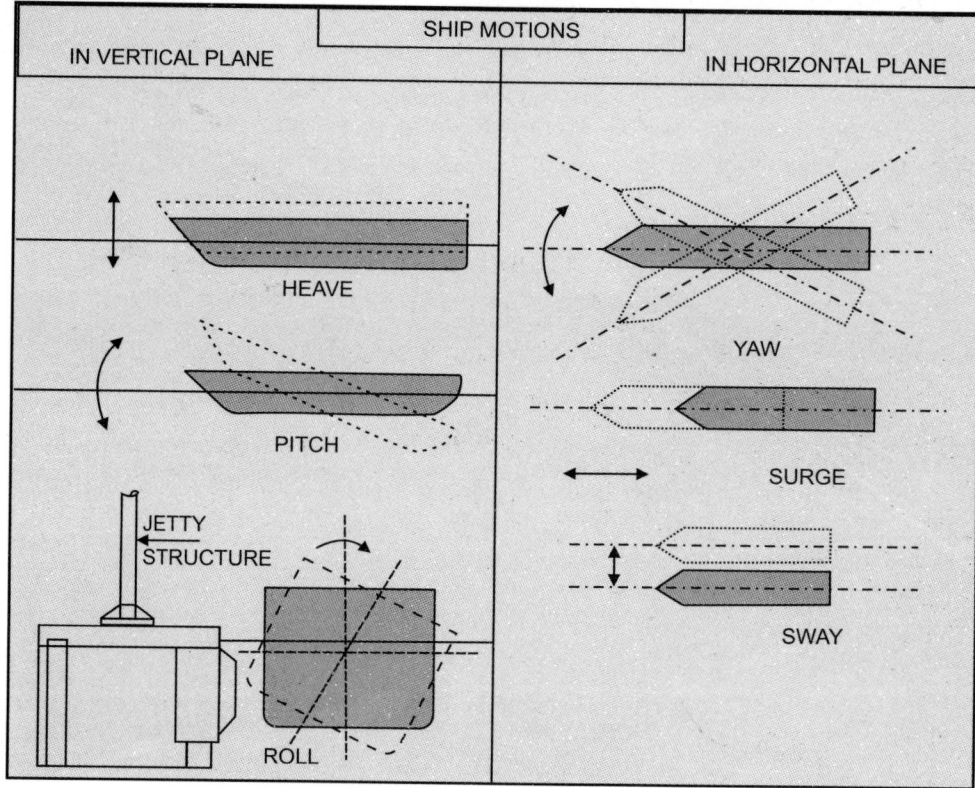

Fig. 5.4: Ship motion in horizontal and vertical planes

on the orientation of the mooring ropes so as to counter act any adverse ship motions that the tanker may face. For clarity, all the ship motions in the horizontal and vertical planes have been shown in Fig. 5.4.

The orientation of the mooring ropes depend on the direction and magnitude of the forces acting on the moored tanker. If strong wind and current are acting beam to the tanker from the land side, the breast lines, would carry almost all the loads and would prevent the sway motions. If the forces act perpendicular to the transverse axis, the stern and bow lines would take the load to prevent the surge motions. If these forces act quartering to the tanker, the yaw motions would be prevented by the spring lines. The wind force against the tanker may either press the tanker against the jetty if wind is blowing from the seaside or push away the tanker from the jetty face if the wind is blowing from the land side. This type of sway may put the tanker in resonating condition which would greatly hamper loading/unloading of the LPG/LNG cargo.

At times the current force becomes critical. If the moored tanker is subjected to a strong currents-either parallel to or acts at an angle, the current may create long periodic ship movements which may not be conducive to loading/unloading of LPG/LNG cargo. The periodic movements however depend on the stiffness and elasticity of the mooring system as well that of the fendering system. In order to reduce the undesirable ship motions, certain operating limiting velocities have been prescribed for wind and current, in excess of which, shut down should be taken.

In so far as the wave induced forces are concerned, normally an LPG/LNG jetty is constructed in protected waters and as such wave disturbance at the jetty site generally is of low order and consequently the wave induced forces on the moored tanker are not appreciable. But when the jetty is constructed in an exposed locality or in a partially protected waters, wave induced force may become significant and a moored tanker may experience considerable, roll, pitch and heave motions which are considered to the most dangerous motions for the tanker. The present day knowledge however is not sufficient to correctly estimate out extent of the adverse motions that would be induced. The most reliable method of predicting the response of a moored tanker to such wave action would be to built a physical model of the harbour in a large scale along with the LPG/LNG jetty and accordingly determine the safe site for the construction of the jetty where these motions would remain within tolerable limits. In this regard, limiting operating criteria for safe working of LPG/LNG tankers are given in "Operating criteria in brief for handling LPG/LNG tankers through a jetty system including operational Rules and Regulations" vide Annexure A.

6

Requirements of Shipping Tugs for Piloting LPG/LNG Tankers

6.1 INTRODUCTION

The safety of navigation in confined water and in port approaches primarily is the responsibility of the pilot in command on board the tanker. He has therefore to make a comprehensive safe passage planning before ETA of the tanker and should decide upon key elements of the navigational plan which should include but not limited to the following:

(a) Safe speed having regard to the manoeuvring characteristics of the tanker restricted by draft, giving due allowance for reduction of draft due to squat and heel effect when turning.

(b) Speed alterations necessary en route to counteract environmental forces.

(c) Identification of course alteration points on the basis of the layout of the port's channel and port chart.

(d) Minimum keel clearance required in critical areas en route especially at the entrance zone.

(e) Contingent passage planning for emergency evacuation from the jetty in event tanker catches fire or there is a fire on board the jetty.

(f) Availability of adequate tug power, pilot launch, mooring launches and performance of these floating crafts including the communication sets.

(g) Adequacy and range of visibility of the existing navigational aids (buoys and shore transit/lead marks).

(h) Maximum weather parameters with regard to wind, wave current and Under Keel Clearance (UKC) as would be acceptable to him for safe navigation and their availability just before commencement of navigation of the tanker.

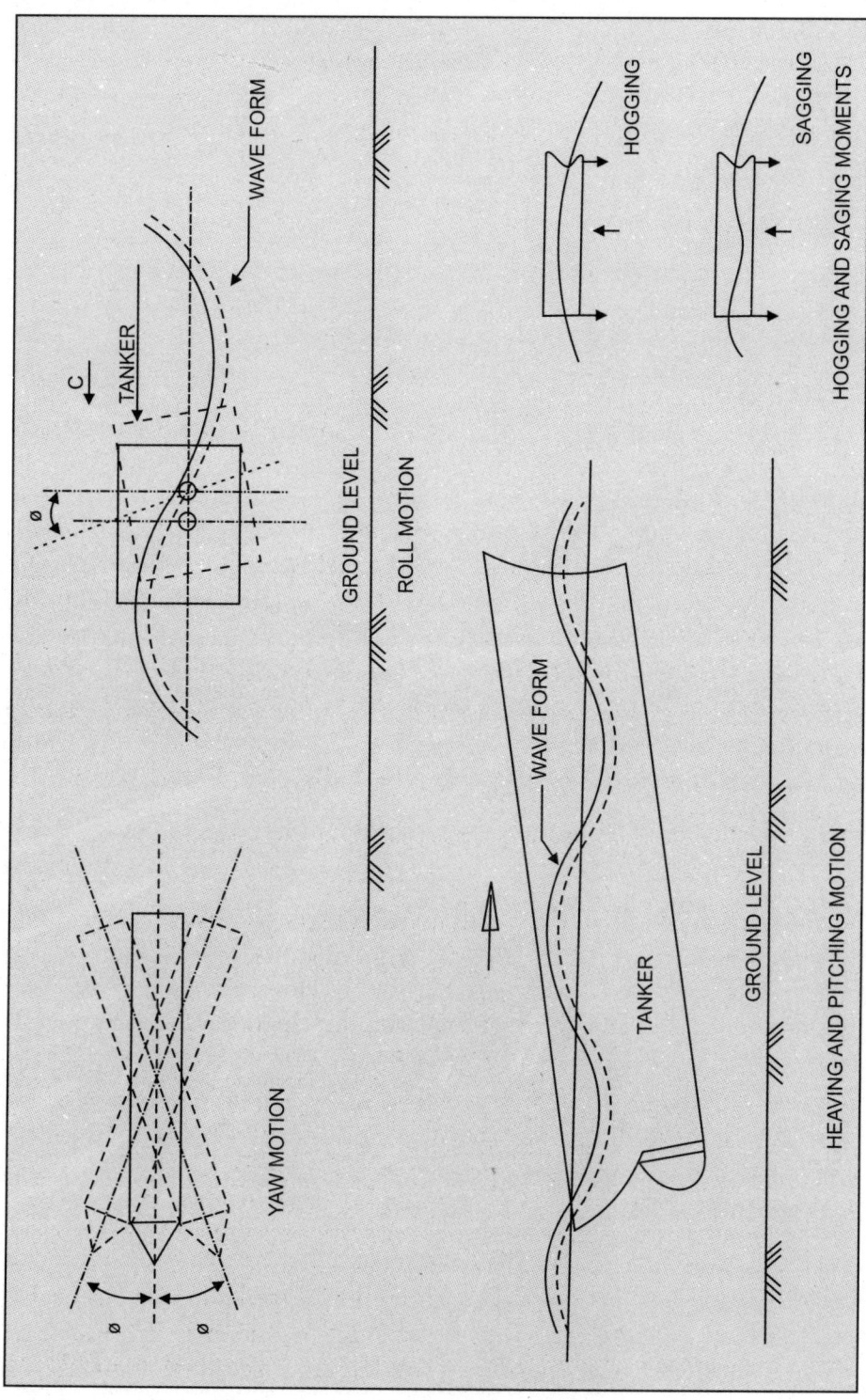

Fig. 6.1: Various ship motions

(i) Acceptable time of navigation (whether it is day light hours or night hours) and additional tidal depths that would be available at that time.

(j) Condition of jetty fenders, quick release hooks, mooring ropes and proper functioning of the fire fighting system including the fire water curtains on board the jetty.

6.2 ASSESSMENT OF TUG POWER

The tug power is assessed in terms of Bollard Pull (BP) and would depend primarily on the size of the tanker, its draft (fully laden or in ballast) and strength, direction of wave, wind and current forces.

6.2.1 Wave Action

In so far as the wave climate is concerned, in beam seas oscillations of sway, heave and roll motions will be felt. In head or stern waves, oscillations of surge, heave and pitch motions will be felt. In quartering seas all the above three motions including yaw motion will occur (Fig. 6.1). If these motions remain within certain tolerable limits, wave field will have negligible influence specially on the navigation of large tankers. However, in higher order wave heights in excess of 1.5 m (Hs) and wave periods in excess of 10 sec. say 14 to 45 sec. periods, there may be three dangers as indicated below:

Streaming into high head seas would cause hogging and sagging moments when the tanker is alternatively supported by waves at her ends and then by a single crest midships (Fig. 6.1). If condition become serious, permanent structural damage would manifest such as cracking of sheer strack and ship's side platting. Indeed high waves have known to have broken ships into two.

In streaming into beam seas, rolling would be the major danger. The major danger in heavy rolling is that of cargo and fittings shifting, leading to vessel taking a list and eventually capsizing. However, with quartering seas rolling would often be at its worst and a fair degree of pitching would also be present.

The greatest danger of all would occur when a tanker is moving in a seaway and tanker's rhythmic movements coincide with the apparent period of the waves. This phenomenon is called synchronism and will result in angle of roll or pitching becoming very large and there may be grave risk of cargo shifting or even machinery breaking away from its mountings. If synchronism occurs or is likely to occur, an alteration of course or speed will destroy the sequence. However, neither alteration of the course nor reduction of speed below a certain navigable speed, would be feasible within the channel geometry, specially when the tanker is negotiating a confined channel.

Again in high seas, pilot may not be able to board the tanker at the fairway buoy to assist to navigate the tanker and also tugs may not be able to take positions in open seas.

Tug power has very little influence on these adverse vertical motions. However, she can assist only to alter the course should the tanker gathers a leeway due to the horizontal motions.

In general, there is therefore no apparent solution to the adverse ship motions. All that can be done is to limit navigation of the tankers when the wave height (Hs) is not more than 1.5 m and period not more than 10 secs. so that the adverse ship motions will remain within tolerable limits.

6.2.2 Wind Action

Wind imparts a hydrostatic drag force on the exposed area of the tanker, its strength varying square of the velocity of the wind. When the tanker is subjected to a beam wind, it may gather a considerable leeway. Course can be made good by either pushing or pulling the tanker by shipping tugs. Owing to the bridge which is generally at the aft and shape factor of the tanker, it is probable that force acting on the tanker's hull may not be uniform in the fore and aft directions which may result in a turning moment (yaw). In order to counter act the yaw, tug assistance would invariably be required. A head wind, although its magnitude may be comparatively small, will give a breaking effect to the tanker and is preferred while navigating a fully laden tanker. A tail wind will have the reverse effect and will add to the tanker's speed which would require a longer stopping distance for a tanker to come to a complete halt position within the TC and is generally not preferred at the arrival manoeuvre.

Gas tankers in general have a high wind area especially when they are in ballast. Consequently they would catch a large amount of wind force. On the contrary, deeply laden tankers have a considerable lesser wind area. Laterally projected wind and current areas above and below sea level for some of the selected LPG/LNG tankers in fully and ballast loaded condition are given in the Table 6.1.

Under the influence of strong beam winds when the tanker is in ballast and when it starts reducing speed, the deflection from the intended track would be proportional to distance steamed along that course. At this stage rudder angle has to be applied to bring the tanker back to the intended course. In most tankers maximum possible rudder angle is 35° starboard or port, although service maximum angle at full speed is voluntarily held at 15° to minimise heeling and strain on the steering gear. If speed of the tanker is slow and keel clearance is low and wind is very strong, rudder may not

Table 6.1 Laterally projected areas for LPG/LNG tankers

Size of tankers (m³)	Above Sea level		Below Sea level	
	Fully loaded (m²)	Ballast loaded (m²)	Fully loaded (m²)	Ballast loaded (m²)
LPG Tankers:				
75,000	2400	3640	2970	1760
52,000	2100	3000	2450	1560
24,000	940	1850	1790	920
LNG Tankers:				
1,25,000	7320	8430	3190	2120
87,600	5900	6780	2490	1660
29,000	3160	3700	1540	1030

respond fully to command even if above angle is given. At this time, course drift has to be reduced with the help of tug power. Even if tug assistance is taken, a minimum speed of 4 knots should be maintained in the tanker to make the rudder functional. However, to avoid a dangerous course drift, navigation should be stopped when wind speed exceeds 20 knots about 40 KMPH (Beaufort scale 5).

6.2.3 Current Action

Like wind, current imparts a hydrostatic force on to the tanker. When she is in fully loaded condition, her exposed area to current would be maximum and when she is in ballast the exposed area would be minimum. Course deviation are greater for lower speeds of a fully laden tanker resulting a wider sweep path. For example in case of a cross current of 0.5 m/sec (1 knot) and when an unassisted tanker is travelling at speed of 4 knots, she has to be steered at a drift angle of 14°. However, this may not be acceptable specially when the tanker is negotiating a restricted channel. Hence, the course has to be made good by using shipping tugs. It is therefore advisable to stop navigation when current velocity is greater than 0.75 m/sec. (1.5 knots) acting abeam, so as to avoid dangerous course drifts.

6.3 ENTRY MANOEUVRES

The manoeuvres of an LPG/LNG tanker should be left to the sole discretion of the pilot in command. However, following manoeuvres may be considered by the pilot handling the tankers:

(a) Attach tugs outside at a convenient point and start assisting the tanker outside of the protective breakwaters so that the tanker can enter at a minimum manoeuvring speed of about 3 knots and then bring the tanker to a complete halt within the TC of the port without any overshooting, by stern power of both the tanker and the tug. Thereafter bring the tanker alongside of the jetty by parallel berthing, stern-in first so that bow faces deep water ahead as a means of evacuation, preferably wind ahead.

(b) Attach the tugs at fairway buoy with long lines and start assisting the tanker only after it enters port under its own power which would require a minimum entry speed of about 4 knots. Repeat the balance operations as indicated in (a).

(c) Bring the tanker under its own power into the outer basin of the port protected by breakwaters and then attach tugs inside to assist which would require a minimum entry speed of 5 to 6 knots. Repeat balance operation as indicated in (a).

According to an experiment conducted by Ashdod Authorities, Israel (Ref.8) for 1,50,000 DWT coal carriers, certain conclusions have been derived on manoeuvring of ships assisted by tugs which can more or less be made applicable for manoeuvring LPG/LNG tankers as indicated below.

In the above three manoeuvres, (a) would require comparatively minimum stopping distance of about 900 m (3 LOA) which however would prove risky for LPG/LNG tankers. Manoeuvre (c) would involve a stopping distance of 3000 m (10–11 LOA) which may be uneconomical. Manoeuvre (b) would involve a stopping distance of about 1500 m (6 LOA) which confirms to IS-4651 Part-V and can be adopted for handling LPG/LNG tankers.

6.4 SERVICEABLE PARAMETERS

In light of the above analysis, safe serviceable navigational parameters may now be fixed for LPG/LNG tankers which would require least tug power. The maximum serviceable environmental parameters perpendicular to long axis of the tanker should be limited to: wind 20 knots (about 40 KMPH), wave ht. (Hs) 1.5 m, wave period 10 secs., current 0.75 m/sec (1.5 knots) and visibility 4000 m. This implies that if available parameters are in excess of these limits, the tanker has to wait at the fairway buoy till the weather improves and attains these levels. Tanker handling in near beam waves is more preferred by many pilots than with quartering waves. As far as possible, in fully laden condition, tanker to face wind ahead. Dia of TC should be 2 × LOA, channel width 5 to 8 Beam or one LOA and stopping distance 6xLOA, reckoned from centre of TC to tip of the protective breakwaters. Navigation of LPG/LNG tankers should be confined to day light hours and in high tide conditions. Since gas dangerous zones extend 24 m or more beyond the side shell of a tanker, unregulated crafts should be prohibited from approaching the tanker on tow.

6.5 TUG POWER

While deciding the tug power, it is desirable to take account of serviceable parameters as covered above. The pilot and the nautical staff may be fully involved in the design process.

6.6 ASSESSMENT OF BOLLARD PULL OF TUGS

Tug power is expressed either in terms of Bollard pull (BP) or Horsepower (HP). The former is more common nowadays. Several studies have been carried out correlating BP or HP of tugs based on size of tankers to be handled, wind and current speeds and initial approach velocity of the tankers.

M/s. Griffith and Hudson made elaborate studies on the tug power in relation to the ship size and presented the results in First International Tug Conference which are given as under:

Table 6.2 Recommended bollard pull of tugs for various sizes of vessels

Size of vessels (DWT)	Total Bollard pull required (Tonnes)	No./composition of Tug power
50,000	60	2 × 30T or 4 × 15 T
100,000	80	4 × 20T
150,000	94	4 × 23.5 T
200,000	105	4 × 26.5 T

The Japanese have evolved standards for different sizes of vessels on basis of the HP which are given below:

Table 6.3 Recommended tug power as per Japanese standards

Size of tankers(DWT)	Tug power required (HP)
30,000	3000 × 2
80,000	3000 × 2 to 3
200,000	3000 × 4
300,000	3000 × 4 to 6

Apart from the above, two guidelines are also available to assess the tug power. One rule defines that for each 800–1000 DWT of the vessel, one BP would be required. The second rule is that each 80–100 HP gives one BP.

BP can also assessed analytically taking account of the wind speed and current speed. The following expressions are proposed for evaluating BP of sea going shipping tug for handling LPG/LNG tankers.

6.6.1 Wind Force

The wind force acting on a tanker can be calculated from the following expression:

$$PW = 0.00481 \times Vw^2 \times Kw \times Kg$$

Where,

Pw = Pressure due to wind in kg/m^2

Vw = Velocity of wind in KMPH

Kw = A constant which can be taken as 1.5

Kg = Gust factor which can be taken as 1.2

6.6.2 Current Force

The current force acting perpendicular on a tanker can be calculated from the following expression:

$$Pc = 8.29 \times Vc^2 \times Kc$$

Where,

Pc = Pressure due to current in kg/m^2

Vc = Velocity of current in m/sec.

Kc = Constant which varies from 1.5 to 6 depending on the ratio of the water depth available to tanker's draft which can be taken as 3.

6.6.3 Total Bollard Pull

The total BP required to move a tanker can roughly be calculated from the following expression:

$$Total\ BP = Sf\ (Pw + Pc)$$
$$= Sf\ (Pw \times Aw + Pc \times Ac)$$

Where,

Aw and Ac = Laterally projected area of the tanker either in fully or in ballast loaded condition for wind above sea level and for current below sea level respectively in sq m.

Due to uneven BP when several tug boats are used and due to some inaccuracy in method of calculating the forces, the BP actually required to move a tanker against these forces should be 30 to 50 % more. Hence the factor Sf can be taken as 1.4.

6.6.4 An Illustration

As an illustration, consider the following fleet of LNG tankers which have to be navigated against the following environmental forces:

• Wind velocity beam to tanker = 10 m/sec. (20 knots; 36 KMPH or say 40 KMPH).

- Current beam to tanker = 0.75 m/sec. (1.5 knots).

On the basis of the above expressions, requirement of shipping tugs works out to as under:

Table 6.4 Requirements of tugs

Size of tanker (m3)	Wind Force (T)	Current Force (T)	Total Force (T)	Bollard Pull (BP)	No. of Tugs Proposed	HP of Engine (HP)
1,25,000	33.24	41.55	74.79	105	Total: 120 BP	
					1 × 40 – 45 BP	1 × 4000
					1 × 30 – 35 BP	1 × 3000
					2 × 20 – 25 BP	1 × 2500
87,600	29.38	34.28	63.66	89.0	Total: 90 BP	
					1 × 35 – BP	1 × 3500
					1 × 30 – BP	1 × 3000
					1 × 25 – BP	2 × 2500
29,000	13.05	25.04	38.06	53.30	Total: 65 BP	
					1 × 25 – BP	1 × 2500
					2 × 20 – BP	2 × 2000

NOTE: All the tugs are tractor tugs.

It is advisable for the ports handling LPG/LNG tankers and large bulk carriers that they should have at least 2–45 BP tugs in their fleet composition so as to give them some reserve capacity in order to handle such category of ships specially in choppy weather that would prevail during monsoonic periods.

6.7 ABILITY OF A SHIPPING TUG

A tanker handling tug must have the ability to manoeuvre rapidly within a confined space and its quality may be judged by the following:

(a) Time taken to complete a cycle of 360° at maximum rudder angle.

(b) Diameter of the turning circle.

(c) Time taken to reach dead stop when running at maximum speed.

(d) Forward, astern and side bollard pulls.

(e) Free running

(f) Thrust control

(g) Behaviour in bad weather

(h) Stability specially in respect of gritting.

The larger the force which a tug is required to transmit to the assisting tanker, the response has to be more precise in order to avoid damage to ropes and mooring gear. When a tanker is coming alongside or getting away, it will involve frequent manoeuvres from ahead to astern and vice versa. Such change over should be smooth when the thrust is passing through zero position. Further, a tug must at all times be fully manoeuvrable even in bad weather or any other cause which reduces the speed. In an emergency it should be even capable of holding itself to sea at zero speed.

The role of the tugs in assisting a tanker should be left to the discretion of the pilot on board the tanker. He can use them fore and aft on short wires or make them to act as thrusters free to move to the position of greatest advantage.

6.8 CHOICE OF PROPULSION SYSTEM FOR TUGS

Propulsion system of tugs broadly consists of Kort Nozzle with fixed types of duct, propellers are fitted to tail shafts and are stern mounted. These propulsion system can give thrust only in two directions, forward or astern. In the right angled steerable propulsion system, to which brand names like Schottel, Rex, Aquo Marine, 'Z' peller, Dock Peller and Harbour Master belong, the propeller rotates at about 300 RPM and the thrust is obtained by combining the power of two propellers vertically mounted to give a thrust of different magnitude and direction. In cyclodical propulsion system, which is manufactured by Voith Schnieder, the blades are mounted vertically and are operated by kinematic lines for providing independent variable pitch which enables thrust to be made available in any direction. The normal configuration with this type of propulsion system is the "tractor tug".

Kort Nozzle tugs can only assist a tanker at speeds below 3 knots because they cannot 'open up' at high speeds. Tractor tugs are much more manoeuvrable, they can be attached at any manoeuvring speed and can assist a VLCC travelling at a speed of 4 to 5 knots. It is important to know under what wave conditions the tug can assist the tanker. The following limiting conditions are accepted (based on experience of South African Ports and elsewhere).

Table 6.5		
Type of tug	Limiting Hs for assisting (M)	Occurrence for 'T' less than 6 secs
Conventional (heavy)	0.5	37.5
Tractor tugs	1.0	25.5
Tractor tugs with rendering winches	1.5	14.1

In view of the above, Kort Nozzle tugs are not in much use in Indian ports. Currently cyclodical system (Voith Schnieder propulsion) is being preferred in Indian ports on account of easy manoeuverability, provision of equal thrust in all directions, faster response to command, effective thrust produced in any direction, simplicity of operations and harder to girt (capsizing).

6.9 TUGS WITH SPECIAL FEATURES

For handling LPG/LNG tankers tugs shall be equipped with Fire Fighting equipment consisting of telescopic monitors with remote control operation and should follow the regulations given below:

External lighting on the weather deck should be flame proof; exposed electric power circuits and switchgear on deck should be explosion proof; the engine room sky light should be gas tight; the ventilation of the engine room and other enclosed spaces should be protected against flammable and toxic gasses, the exhaust from the main engine should be fitted with spark arresters and tug to be fitted with a dual frequency echo sounder.

Hazard Analysis

7.1 INTRODUCTION

When liquid flammable gas escapes, approximately 30% (depending on the boiling point and degree of pressurisation), will flash to form a vapour cloud. If it finds a source of ignition before being dispersed below the Lower Flammability Limit (LFL), a fire may occur and the flame may travel back to the source of leak. Any person caught in the fire is likely to suffer fatal burn injuries. Such a phenomenon accompanied by explosion, may occur at an LPG/LNG jetty system. Apart from this, such hazards may occur when a ship gets grounded or she collides and rams with another stationery or moving ship.

7.2 DEFINITIONS

LNG: This is the abbreviation for the liquefied natural gas predominantly containing methane. LNG is obtained by liquification of natural gas which contains methane, small quantities of other hydrocarbons (collectively known as NGLs), varying amount of water, carbon dioxides, nitrogen and other non-carbon substances. LNG can be liquuified at (–) 161.5° C and can be transported either in pipelines as a gas or by sea in the liquid form.

LPG: This is the abbreviation for liquefied petroleum gas which can be produced from NGLs or from refining crude. The constituent elements are propane or butane or a combination of both. Propane can be liquefied at (–) 42.3° C and butane at (–) 0.5° C and can be transported in that form in ambient temperatures.

BLEVE: Boiling Liquid Expanding Vapour Explosion phenomenon can occur when a large volume of flammable vapour is suddenly released from a pressure vessel in presence of a source of ignition and the vapour cloud which is nearly spherical may burn as a fireball accompanied by explosion. The heat radiation and blast over pressure from such a fireball are very intense and can cause a great deal of damage.

BOILING POINT: The temperature of a liquid at which a liquid changes its state from a liquid into gas or vapour at a given pressure. For water at sea level boiling point is 100°C.

BOIL OFF: The vapour produced above the cargo liquid surface due to ingress of external heat into the cargo.

EXPLOSIVE RANGE: The range of concentration of a flammable gas or vapour (% by volume of air) in which explosion can occur upon ignition in a confined area.

FLASH POINT: The lowest temperature at which a liquid gives off a vapour sufficient to form an flammable mixture with the air near the surface of the liquid. Flash point temperature is determined by laboratory in a prescribed apparatus.

HARD ARM: An articulated metal arm used at the terminal jetties to connect shore pipelines to the ship's manifold. This is also termed as the metallic unloading arm/marine hard arm.

MANIFOLD VALVES: Valves in a tanker immediately adjacent to the ship/ shore connecting flanges.

SLOSHING: The splashing of cargo into cargo tank due to the ship movement resulting from the wave effect. It is enhanced if free surfaces are present.

BUND/DYKE: Raised ground or wall built around a tank, tank farm, hard arm to retain spills and prevent their spread, thus reducing the hazard.

FLASH FIRE: Burning of flammable vapour at low flame propagation speed, low enough for expansion to take place easily with significant overpressure ahead or behind the flame front. The hazard is only confined to thermal effects.

POOL FIRE: A pool of flammable liquid burning with a stationery diffusion flame.

7.3 INTRINSIC HAZARDS OF LPG/LNG FACILITIES

Safety is particularly an important aspect while ship to shore transfer or viceversa of liquefied gases. In event of a fire, there would be radiation of heat from the flame zone and in event of an explosion, blast over pressure would develop. Both may cause devastating damages.

7.4 FLAMMABLE AND EXPLOSIVE RANGE

The range of gas concentration in air has an important bearing with the flammability and explosive characteristics of the released gas. This describes the range of concentration between LFL (Lower flammable limit) or LEL (Low explosive limit) and UPL (Upper flammable limit) or UEL (Upper explosive limit). Vapour mixtures within these ranges are capable of being ignited resulting a fire and can also explode. These ranges has been shown in Fig. 7.1.

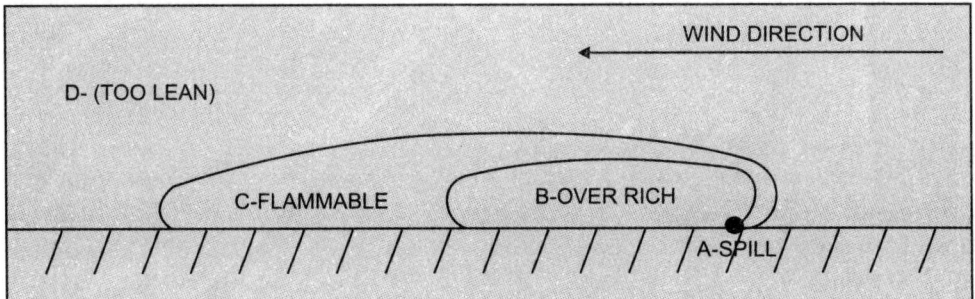

Fig. 7.1: Flammable vapour zones, a liquefied gas spill

The mixtures of below LFL is too lean and above UFL is too rich and therefore are not flammable. All the liquefied gases with the exception of clorine are flammable but limits of flammable range vary depending on the particular vapour. These are listed in Table 7.1.

Table 7.1 Ignition properties for liquefied gases			
Liquefied gas	**Flash point**	**Flammable range (% by vol. in air)**	**Auto-ignition temperature (°C)**
Methane	− 175	5.3–14	595
Ethane	− 125	3–12.5	510
Propane	− 106	2.1–9.5	468
η-Butane	− 60	1.5–9.0	365
η-Butane	− 76	1.5–9.0	500
Ethylene	− 150	3–34.0	453
Propylene	− 108	2–11.1	453
α-Butylene	− 80	1.6–10	440
β-Butylene	− 72	1.6–10	465
Butadiene	− 60	1.1–12.5	418
Vinyl chloride	− 78	4– 33	472
Ammonia	− 57	14–28	615
Chlorine	NON FLAMMABLE		

7.5 Liquefied GAS FIRES

Broadly liquefied gas fires can be classified as under:

7.5.1 Jet Fires

Fire ball or BLEVE, small confined leaks from pump glands, pipe flanges, etc., will initially produce vapours but will not ignite spontaneously. However, large unconfined leaks from pipelines which are likely under pressure would immediately form a vapour cloud and if it meets a source of ignition, this will give rise to a jet flame. Since vapour cloud would be small and would be in a total unconfined state, chances of a vapour explosion is remote. Further this fire may not cause direct damage but because of heat radiation, adjacent equipment, pipes, structures may weaken which may rapture or fail causing further release of gas. If emergency shutdown are not taken and the fire is not extinguished, there is a high risk of further vapour cloud production and the fire may flash back causing reignition.

7.5.2 Pool Fires

A liquid spillage from a pipeline or hard arm rapture may involve a very large quantity of spill and may form a pool in the bunded area over the jetty deck. The dimension of the pool would depend on the quantity of release, the slope of the deck and bund size used for containment. The spill will start vapourising immediately due to transfer of heat from the concrete deck of the liquid pool. If the liquid pool does not catch fire, then the vapour cloud would drift downwind of the pool. The vapour would be cold and heavier than air and would therefore stay near ground levels until it has become sufficiently diluted such that normal atmospheric turbulence will carry it away. The size of vapour cloud is dependant on many factors such as pool mass, surface area, rate of heat transfer to the pool, terrin conditions and atmospheric stability. If air mixture in the vapour reaches a percentage between about 2 to 10%, the vapour will form a flammable mixture and if it meets a source of ignition in its travel path, the vapour cloud will ignite and it will rapidly be consumed and flashed back to the pool area where flames will evolve over the flames. The heat from the flames will be radiated back into the surface of the pool, thus increasing rate of vapourisation of the liquid. These vapours will feed the flames and increase the flame height until a maximum height is reached when pool reaches equilibrium. When fuel feeding the pool is shut off, the pool will start decreasing in size until the flammable fuel in the pool is completely exhausted. The vapour cloud being totally unconfined, chances of explosion are low. But if the vapour cloud is very large, changes of explosion after ignition are high.

7.5.3 Flame Heights

In still air conditions a flame of pool diameter 'D' will attain a height of 2D to 3D. In wind a velocity of say 2 m/s the flame would deflect by about 45° and in a wind speed of 7 m/s, the deflection would be 30°. These fire configurations have been shown in Fig. 7.2. The heat generated within the flame zone is almost instantaneous with ignition. This heat may cause increased leakage from the flanges, valves, spindles and pump glands unless some form of a cooling is undertaken. This also is applied to exposed parts of cargo containment system.

7.5.4 Heat Radiation Zones

The temperature at the base zone of the flame may be the order of 1000–1200°C which may taper down to 800–600°C at a height of 10 m and 100° C at a height of 16 m as shown in Fig. 7.3. Approximately one-third of the total heat generated by the fire will be the radiated heat. The heat flux emitted at the surface of the flame envelope of LNG and refrigerated liquid propane cloud fires is of the order of 170 kilowatts per sqm. The radiation flux received by an object varies approximately as the inverse square of the distance from the adjacent surface of the flame envelope and the receiving object. A human being can stand a radiation flux of 4.5 kilowatts/sqm, which is the threshold of pain. At 6.2 kilowatt/sqm., severe blistering takes place in seconds and at a flux 12.6 kilowatt/sqm. vapourisation of PVC cable protection takes place. Table 7.2 shows damage due to incident radiation intensity.

Table 7.2 Damaged due to incident radiation intensity	
Incident Radiation Intensity: KW/m²	**Types of Damage**
37.5	Sufficient to cause damage to process equipment.
25.0	Minimum energy required to ignite wood at infinitely long exposer (non piloted)
12.5	Minimum energy required for piloted ignition of woods melting plastic tubes, etc.
12.6	Vapourisation of cable protection takes place
6.2	Severe blistering takes place
4.6	Sufficient to cause pain to personnel if unable to reach over within 20 seconds. However blistering of skin (first degree damage burns) is likely
1.6	Will cause no discomfort to long exposer
0.7	Equivalent to solar radiation.

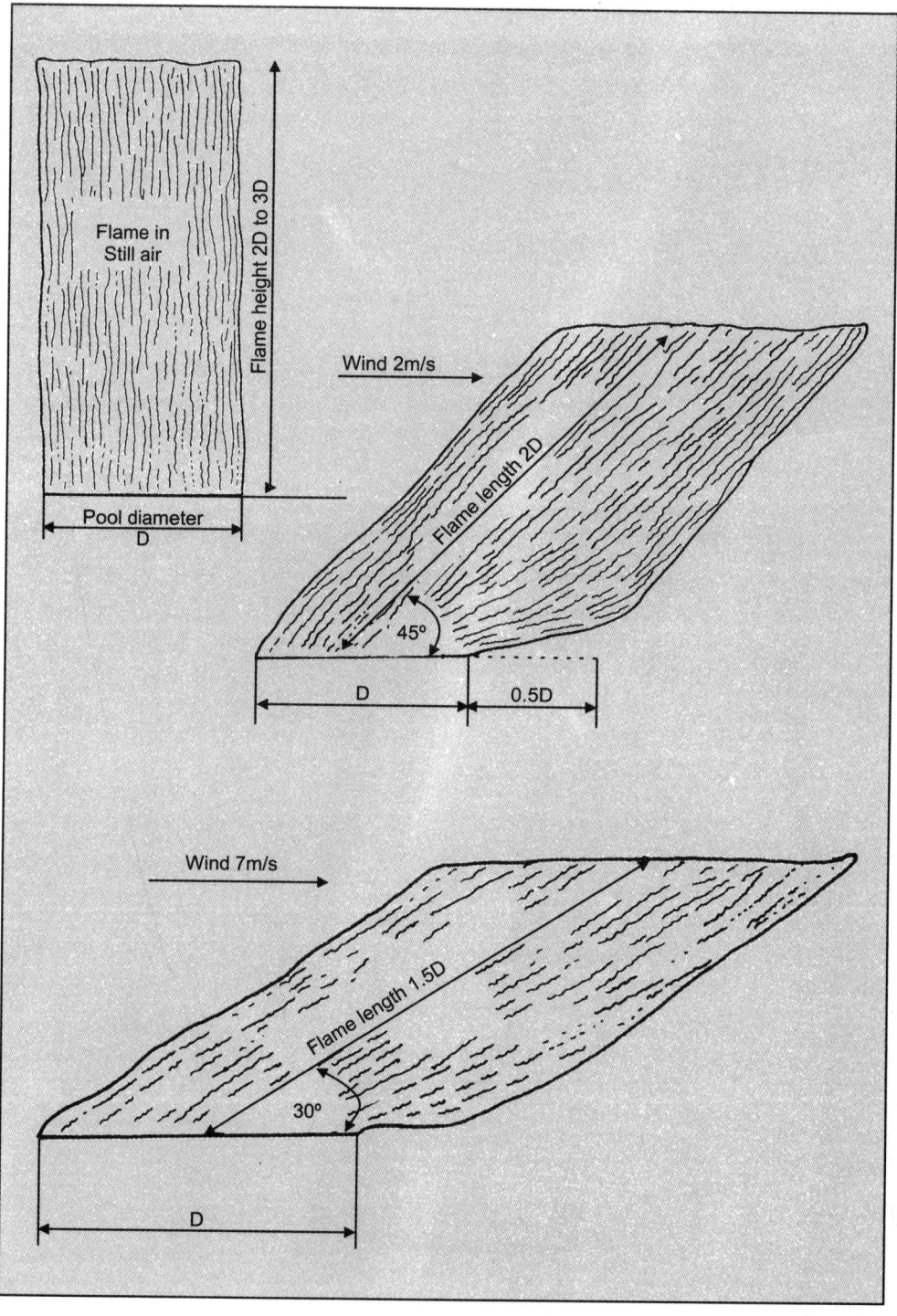

Fig. 7.2: Pool fire configuration

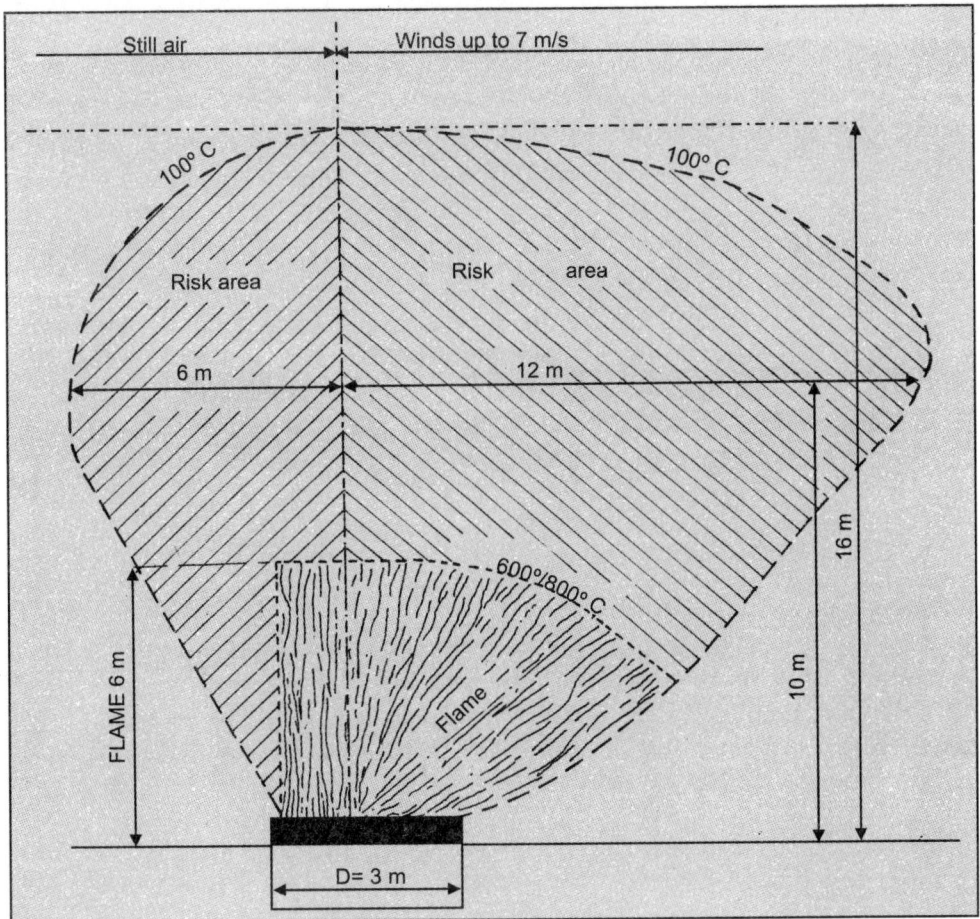

Fig. 7.3: Behaviour of the flame

LNG vapours burn with comparatively clear flame; LPG burns with a greater production of shoots and as a result, maximum surface emissive powers are lower than for LNG. Based on the Maplin sand tests, the surface emissive powers may be of the following order:

Material	Surface emissive powers	
	KW/m²	
	Cloud fire	Pool fire
	173 ± 25	203 ± 31
Refrigerated liquid propane	173 ± 20	43 ± 9

Heat radiation both from LPG/LNG fires dictate that unprotected personel must leave from the immediate vicinity of the pool fire as quickly as possible.

7.5.5 Liquefied Gas Explosions

Consequent to the accidental release of liquid LPG/LNG, the vapour cloud may ignite and explode causing high blast over pressures and consequently may cause very heavy damages. If explosion occurs in a unconfined condition it is termed as percussive unconfined vapour cloud explosion (PUVCE). Even though large quantities of LPG/LNG emission is necessary, only a fraction contributes to the percussive effect. Rare though, PUVCE may be, their damage may be large and sometimes enormous.

If the explosion occurs in a confined condition, damages will be extremely severe. If a flammable mixture contains adequate amount of oxygen and meets a source of ignition, combustion occurs, by which process hot gases are released. In a confined space if expansion of the hot gases is restricted, pressure rises and the speed of flame travel increases. This depends on degree of confinement encountered. Increased flame speed gives rise to more rapid increase in pressure with the result that an explosion may occur with damaging over pressure may be produced. In partially confined conditions such as a surrounding pipe work, plant and building, explosion may take place resulting undue over pressure. However, due to turbulence in the air, vapour will rapidly mix with air, making it an over rich mixture and therefore in an unconfined or a partially confined situation chances, of an explosion are rare. However, damage effects of blast over pressure are given in Table 7.3.

Table 7.3 Damage effects of blast over pressure	
Blast over Pr (Psi)	**Damage level**
5.0	Major structural damage (assumed fetal to people inside the building or within other structures)
3.0	Oil storage tank failure
2.5	Eardrum rapture
2.0	Repairable damage, pressure vessel remain in tact light structures collapse
1.0	Wind breakage, possible causing some injuries.

7.6 BLEVE

In confined conditions a phenomenon called BLEVE (Boiling Liquid Expanding Vapour Explosion) may occur which is very devastating. If the critical wall of containment such as a pressure vessel containing pressurised LPG is subjected to a flame impingement due to an external source, the temperature inside the container increases and consequently the internal pressure also increases. The temperature increase also weakens the welded joints of the container to the point of failure. As a result when there is sufficient built up of pressure, then due to the blast, pieces of vessel's shell would be thrown considerable distances with concave sections such as end caps being propelled as rockets. Upon rapture, the sudden decompression produces a blast and the pressure immediately drops. At this time the liquid temperature is well above its atmospheric boiling point and accordingly it spontaneously boils off, creating large quantities of vapour which are thrown upwards along with liquid drop lets. The released clouds of vapour burns as a 'fire ball'. This is nearly a spherical cloud of flammable material burning with much turbulence and rising as it mixes with the surrounding air. In this case the blast over pressure and heat radiation are very intense and can cause intensive damages.

BLEVE incidents have occurred in rail tank cars, road vehicles and in a number of terminal incidents. This hazard however is unlikely to occur at a jetty site since the storage area is situated at a far off place uplands. At a jetty with hard arms confined by a concrete dyke, the spillage of the hydro-carbons would be of low order as the source of spill will be generally from flange joints or bleeder valve left open inadvertently. However, maximum credible accident can be the failure of hard arm or full bore failure of a hydro carbon gas pipe.

In that event, large quantity of LPG/LNG may spill and form a pool inside the bundled area and eventually pool fire may result. Since the pool space is partially confined chances of an explosion occurring would be remote.

7.7 SOURCES OF IGNITION

Accidental sources of ignition can be flames, thermal sparks (due to metal-to-metal impacts) and electrical arcs or spares. Some of the sources could be from smoking, hot welding, lighters, matches, torches, domestic equipment like shoes with studs and nails, shavers, radios, electric cooking appliances, hand tools, electric tools, mobile telephone, electricity sparks arising out of connecting and disconnecting cargo connections between ship and shore and short circuiting of electric cables.

7.8 HAZARD ANALYSIS FOR THE JETTY STRUCTURE

The hazard analysis gives three basic results

- For estimating LFL zone and blast over pressure zone and accordingly all the vital installations can be located at safe distances.
- For arriving at the optimum location of the proposed jetty which will produce minimum damage in event of an accident.
- For evolving mitigating measures.

7.9 APPROACH TO THE ANALYSIS

In so far as a jetty is concerned, small spillages from pipe/valve flanges would not be of any credible consequence. Full bore failure of a pipeline and failure of an unloading hard arm will constitute as the maximum credible accident. It is however recognised that failure of a large pipeline is of rather infrequent occurrence. But the very nature of operation of the hard arm makes it more vulnerable to fracture than a single piece of pipeline. The frequent connections, disconnections, swivel joint motions all around, relative movements between the jetty and the alongside ship in adverse weather strained conditions make the swivel joints more vulnerable to failure. Moreover the adverse motions like surge, sway, heave and roll motions which the ship may experience while alongside the jetty due to poor tranquillity conditions and in slack mooring conditions or parting of mooring lines and in elevation changes for tide and draft, the hard arm may go out of the prescribed operating envelope. Under this condition, swivel joints may be strained to their breaking point. Hence failure of a swivel joint although may not be frequent occurrence, may be realised during life time of operation. The use of flexible hose instead of an hard arm increases the failure chances to a credible limit. Whereas in case of a hard arm, it is possible to take an immediate shut down in adverse conditions and disconnect it from ship's manifold by an automatic device, but in a flexible hose system, however, such arrangements are not possible. Hence failure chances are more in case of a flexible hose than a hard arm. In light of the foregoing discussion it can reasonably be concluded that full bore failure of the hard arm can be considered as the maximum credible accident in a jetty system.

In even of failure of the hard arm, there may be significant spillage of LPG/LNG before shut downs are taken. This spill may fall on the jetty surface or on the deck of the ship from where it may spill down to sea water. Usually the jetty area where hard arms are operating is bunded with a concrete dyke. The spill on the jetty surface will remain confined to this bunded area. It may be mentioned that without this confinement by a dyke wall, the spill on the jetty is likely run off over the edge into sea water. As indicated earlier, the pool would almost start evaporating to form a vapour cloud and on mixing with air will become flammable and will move downwind. If it meets a source of ignition on its way, it will get ignited and may explode and may travel back to the source of leak. This phenomenon may also occur on the jetty deck. In either of the cases, the resultant blast waves and thermal radiation may have serious damaging effects.

If the spill falls on sea water, vapour cloud will form more rapidly as compared to the spill on the concrete deck and will as usual drift downwind and if it finds a ignition source before it safely disperses, it will ignite. In case of a delayed ignition, the flammable cloud may cause a vapour cloud expansion and may generate blast over pressure.

If cold LNG liquid spills onto the ships deck, not designed for low temperature, it may chill the steel to a temperature where it becomes brittle. Stress already within the steel together that resulting from differential contraction, can cause fracture in the cooled area. However to avoid these occurrences, in all the refrigerated gas carriers in the area around the manifold, stainless steel, wooden or equivalent drip under manifold connections, are provided.

7.10 WEATHER CONDITIONS

Dispersion of vapour cloud is a wind-borne phenomenon. The flammable hazard is in the direction of wind. In near still wind condition, vapour cloud may be very slow to disperse. In high winds, the vapour cloud may travel a long distance and due to atmospheric turbulence may disperse without any ignition or explosion. In pasquill stability of atmospheric classification, class 'A' is one having most unstable weather conditions. An unstable weather condition promotes better dispersion. A high wind speed and high solar radiation favour the formation of unstable weather conditions. Turbulence induced by buoyancy forces in the atmosphere is closely related to the vertical temperature gradient. When the temperature decreases with height at a rate of 1 degree/c, per 100 m [Adiabatic Lapse Rate (ALR)], the atmosphere is deemed to be in neutral condition (Posquill category-D). If ALR is more than this value, the atmosphere is said to be unstable and categorisation moves towards C, B or A. With lower ALR value the atmosphere tends to become stable (towards F). Vertical motion in the atmosphere is enhanced when ALR is high, i.e., in unstable conditions; dispersion is facilitated. When temperature decreases at a lower rate, vertical motions are dampened or reduced; dispersion is adversely affected under such conditions.

7.11 ESTIMATION OF DISPERSION DISTANCES

The estimation of dispersion distances involve calculation of the following:
- Lower flammability levels (LFL).
- Blast overpressure distances.

These calculations are made under the following broad assumptions:
- Rate of discharge.
- Posquill category (unstable, neutral or stable).
- Wind velocity at the site round the year and its direction at the jetty, round the year.
- Terrin conditions.
- Duration of release.
- Pressure and temperature of the cargo.
- Atmospheric temperature.

Correlating these parameters and other site specific parameters, several correlations, monographs, worked out examples, fire and explosion calculations are carried out. TNO's yellow book sets out elaborate methodology which are used for related calculations. For explosion calculations DM method as indicated in the same book is followed. These correlations have now been back coded into computer programming such as HEGADAS model II by Shell International which facilitate fast and error-free results. However estimation of dispersion distance in line with the above methodology is outside the preview of this book.

7.12 CASE STUDY I : HAZARD ANALYSIS OF LPG JETTY AT PIPAVAV PORT (Gujarat)

The design and hazard analysis of the proposed LPG jetty and obtaining approval of CCoE Nagpur of Pipavav Port (Gujarat), designed to service 60,000 DWT ships (pressurised/refrigerated LPG) was awarded to the author's firm (United Consultants). Hazard analysis got conducted for handling pressurised and refrigerate LPG through an associated Consultant, Triune Ltd., New Delhi. The results of this study with respect to blast over pressure and LFL distances are given in Table 7.4.

Table 7.4 Flammable and blast over pressure distances for unloading arm failure (pressurised LPG)

Release Rate	Wind Speed	Weather Stability	LFL Distances	Blast over pressure distances (M)		
(t/hr)	(m/s)	Posquill category	(M)	5 psi	2 psi	1 psi
150	7	D	113	40	81	203
150	3	D	176	66	132	330
200	2	F	378	100	201	503
200	7	D	128	43	87	219
200	3	D	207	66	113	333
250	2	F	443	101	203	509
250	7	D	140	55	110	275
250	3	D	232	87	174	435
250	2	F	498	118	237	594

Spill duration: 3 minutes: Spill quantity: 97 kg for 3 minutes duration: Ambient temp: 42°C: Max. credible accident: failure of hard arm.

The above LFL and blast over pressure distances as tabulated in Table 7.4 and have also been shown in Fig. 7.4. The location of the jetty was

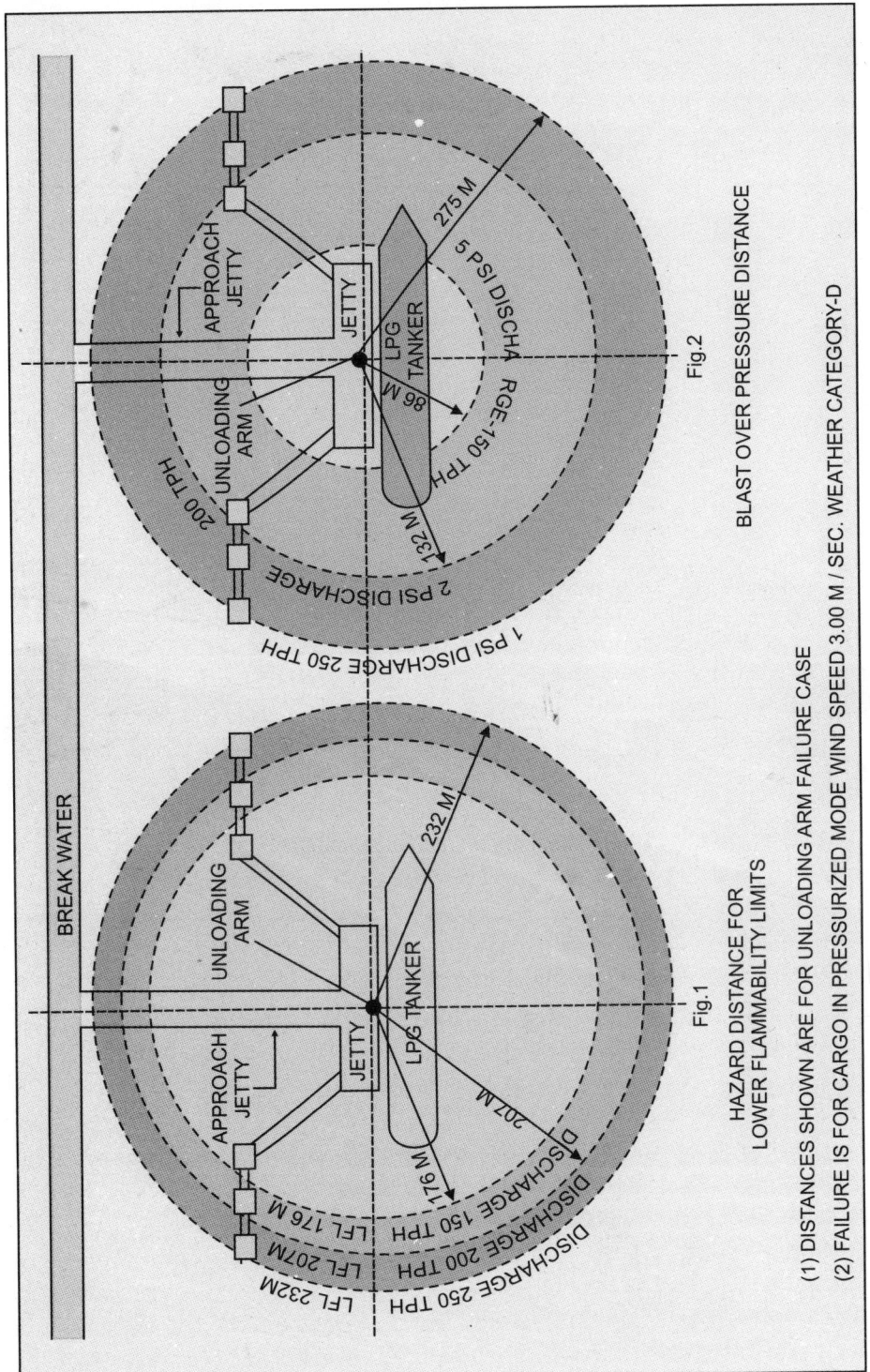

Fig. 7.4: Results of risk analysis for LPG jetty of Gujarat pipavav port

Table 7.5 Comparison of observed and predicted dispersion distances to the lower flammability limit

Material & Spill type	Av. Spill Rate	Av. Wind Speed	Observed Dispersion Distance (3 s average)	Predicted Dispersion Distance HEDGAS III
	m³/m	(m/s)	(m)	(m)
Propane (continuous)	2.3	5.5	215 ± 20 (Peak)	245
	2.8	8.1	245 ± 35 (Peak)	140-225
	3.9	5.6	340 ± 8 (Peak)	320-540
	4.3	7.9	210 ± 50 (Peak)	280
	2.3	3.8	400 ± 100 (Mean)	295
LNG (continuous)	2.9	3.9	150 ± 30 (Peak)	245-335
	4.7	4.5	130 ± 20 (Peak)	235-450
	2.5	4.8	110 ± 30 (Peak)	235

Based on average roughness value

so fixed that all the vital installations, schools, hospitals and other areas of human concentrations, navigational channel, turning circle, cargo berths, etc., fell outside of the maximum distances indicated in Table 7.4. The proposal was approved by Chief Controller of Explosives (CCoE), Govt. of India, Nagpur.

7.13 CASE STUDY II : MAPLIN SANDS OF S.E. ENGLAND

In 1980 at a coastal site in SE England known as Maplin Sands, an experiment was conducted involving spilling of 20 m³ of refrigerated gas and dispersion experiments were carried by Shell International, results of this commendable physical experiment are presented in Table 7.5.

It would be evident that dispersion distances in respect of continuous spill of propane are larger as compared to continuous LNG spill.

7.14 SHIP TO SHORE (STS) TRANSFER OF LPG/LNG

Ship to Shore transfer of LPG cargos has been allowed in many International port terminals. This procedure is carried out in accordance with a comprehensive code of practice such as "Ship to Shore Transfer of "Liquefied Gases", International Maritime Forum". However in Indian major ports, due to safety reasons, STS transfer of LPG/LNG is not a commercial practice. On two occasions as a means of emergency rescue of damaged LNG tankers, this has been permitted in some foreign ports. If, however, STS transfer of LPG is allowed, in Indian Ports as an emergency measure, this may be limited to pressurised LPG and that calm water (wave disturbance not to exceed 30 cm. Wave period 10 sec. and wind 10 knots) is essential with a good fendering system so that the mothership and its lightening ship may quietly work together. Furthermore, it should be ensured that human concentrations, navigational channel, turning circle and port installations are located at least 1500 m away from the STS transfer site. Purpose of the limitation in operating wave height and wave period, is to ensure that the flexible hose discharging the LPG cargo does not get snapped and if otherwise, there would be a significant spill of LPG which will quickly form a flammable vapour cloud and may get ignited and travel back to the source or may explode. In either of the cases, high intensity thermal radiation and blast over pressures may occur which may be very damaging not only to the two ships but also to the working personnel.

STS transfer of LNG by any Indian Ports, for the present, may not be permitted.

ANNEXURE A

OPERATING CRITERIA IN BRIEF FOR HANDLING LPG/LNG TANKERS THROUGH A JETTY SYSTEM INCLUDING OPERATIONAL RULES AND REGULATIONS

In light of the discussions made in various chapters of this book, a consolidated operating criteria in brief for handling LPG/LNG through a jetty system have been furnished in Annexure A for guidance. Majority of these guidelines are also applicable to crude/POL tankers.

Sr. No.	Description	Operating criteria
1	**Navigational requirements**	
	– Dia of turning circle	2 × LOA
	– Width of navigational channel	(a) 5 × Beam in normal conditions (b) 7-8 × Beam (where yaw is experienced)
	– Stopping distance reckoned from centre of TC to the tip of breakwater	6 × LOA
	– Min turning radius at bends	6 × LOA
2	**Draft requirement**	
	In open sea conditions	1.2 × Max. Draft
	In semi protected channels	1.15 × Max. Draft
	In fully protected Basin	1.10 × Max. Draft
3	**Allowable significant wave heights at the jetty**	
	– Wind speed 15–20 knots wave Heading 22.5°	1.5 M/10 sec.
	– Wind speed 15–20 knots wave Heading 45°	1.0 M/14 sec.
	– Wind speed 25 knots wave Heading 45°	0.75 M/14 sec.
	– Wind speed 30 knots wave Heading 90°	0.50 M/14 sec.
4	**Limiting operating wind velocity**	
	– Manoeuvring of gas tankers	10 M/Sec (20 Knots)

Sr. No.	Description	Operating criteria
	– Loading/unloading operation	15 M/Sec (30 knots)
	– Disconnection of unloading arm	20 M/Sec (40 knots)
5	**Limiting operating current velocity**	
	– Normal operation	2 knots (cross current)
	– Stop navigation	4 knots (cross current)
6	**Limiting operating ship motions**	
	Max. Ship motion criteria permissible at the jetty site	Surge - 1 M Yaw - 5° Roll - 3° Sway - 0.5 M Heave - 0.3 M
7	**Limiting visibility**	
	Minimum Visibility	2000 M
8	**Minimum clearances**	
	– jetty face to edge of navigational channel and T.C. for LPG/LNG tankers	500 M or 1 psi Blast Over pressure zone or LFL zone whichever is greater for LPG/LNG tankers.
	– Clear distance between an LPG/LNG tanker and a moored vessel	500 M or 1 psi Blast Over pressure zone or LFL zone whichever is greater.
	– Optimum length of pipeline from an LPG/LNG jetty to Tank farm	2000 M
9	**Requirements of shipping tugs for handling lpg/lng tankers**	
	Upto 29,000 M³	**3 TUGS – 65 BP** 1 of 25 BP + 2 of 20 BP
	Upto 87,600 M³	**3 TUGS – 90 BP** 1 of 35 BP + 1 of 30 BP + 1 of 25 BP
	Upto 1,25,000 M³	**4 TUGS – 120 BP** 1 of 40 - 45 BP +

Sr. No.	Description	Operating criteria
	(To be finalized in consultation with the Port's Pilot)	1 of 30–35 BP + 2 of 20–25 BP
10	**Stand-by tug**	One tug to be kept at a dedicated jetty near to LPG/LNG jetty to pull the tanker off the jetty in the event of fire. The tug shall not be double banked to the LPG/LNG tanker.
11	**Mooring**	– Mooring lines shall be symmetrically arranged. – Breast line to be oriented parallel to the transverse centre line of the tanker. – Spring lines are to be provided parallel to longitudinal axis as far as possible. – Howsers of some material and elasticity shall be used. – Mooring angles as suggested in Chapter-5 are to be adopted – Length of mooring ropes preferably not to exceed 50 m. – Mooring Dolphins are to be placed about 35–50 m off the berthing face of the jetty. – Quick release hooks should be provided.

Other important operational rules and regulations

1. No ship should overtake and bypass an LPG/LNG tanker.

2. While an LPG/LNG tanker is in transit, other ships should vacate the common navigational channels and the turning circle (T.C.).

3. LPG/LNG tankers are to be admitted only during day light hours even if night navigational facilities are available in a Port.

4. While berthing an LPG/LNG, crude and POL tanker alongside a jetty, bow to face deep water, preferably wind ahead, parallel berthing system shall be followed. Angle of approach shall be around 3° and in no case should exceed 5°. Limit transits are to be erected about 40 m off the berthing face of the jetty.

5. While an LPG/LNG carrier is passing, all cargo ships along other jetties on way shall be on taught moorings and a minimum clearance

of 3 × Beam of the largest vessel shall be ensured between the moored vessel and the passing vessel (lPG/LNG tanker) at the entrance zone of the port.

6. Before opening the port to regular shipping of gas, crude and POL tankers the navigational and communicational system (Buoys and T-marks) should be upgraded. All day marks/lights (preferably gas lit) should have a visibility range of not less than 6 NM. All isolated dangers, rock out crops, shallow patches and sunken ships along the shipping channel are to be identified and marked with lights and day marks.

7. Tankers are to be manoeuvred into the harbour preferably in high tide conditions to obtain some extra UKC for better propeller efficiency and better controllability.

8. Safe and protected anchorage area near to the harbour should be identified and marked for anchoring the waiting tankers/disabled tankers exclusively.

9. Deputy Conservator (DC)/Harbour Master (HM) of the Port should inspect an LNG/LPG carrier at the anchorage prior to according permission to enter into the port waters and the Master of the tanker must be prepared to demonstrate that caryogenic handling equipment are in proper condition, failing which DC will refuse to accord permission of entry of the gas carrier into the harbour waters.

10. All foreign flag carriers gas, crude and POL, should carry a letter of compliance of US Coast Guards Regulations or IMO fitness certificate without which DC shall refuse to accord permission of entry to the gas carrier into the harbour waters.

11. No other cargo transfer operation or vessel movements shall be permitted at the jetty site during LPG/LNG transfer operations.

12. No welding, burning, hot work, open lights or similar activities shall be permitted within a radius of 1.5 km while an LPG/LNG tanker is moored at the jetty.

13. DC of each port where LPG/LNG is being handled or contemplated to be handled shall frame LPG/LNG carrier safety regulations, terminal regulations passage regulations and he should promulgate these within his jurisdictional boundary, which among others, should cover prevention of fire on board the jetty and the ship, fire fighting equipment, fire extinguishing and emergency procedures and evacuation plan in case of fire, collision, grounding, disablement, cargo

spills or leaks and personnel casualty including regulation specific to discharge of dirty ballast, bunker and STS (Ship to Ship) transfer of flammable cargo (in case of disablement). This will be applicable for crude and POL tankers also.

14. Control of an LPG/LNG carrier should begin at least 72 hrs of the advance notice of arrival after due inspection. As said earlier, navigation shall be permitted during day light hours and only during the period of good visibility and must await arrival of a pilot's escort vessel and attendance of at least by three tugs for LPG/LNG carriers.

The escort vessel will broadcast frequent radio alerts also by continuous hooting of siren throughout the harbour transit in order to avoid crossing, meeting and overtaking situation involving the LPG/LNG carrier and any other vessel in harbour.

15. STS (Ship to Ship) of transfer of LNG shall not be permitted. Under 10 knots of wind, 0.30 M, Hs, 6–10 sec. wave period, 1 knot cross current. STS, outside of the port limit for transfer of LPG (pressurised form only) may be permitted only as an emergency evacuation plan, after making an STS transfer guidelines for the port in accordance with International Conventions/Guidelines.

16. If the jetty is located near the tip of a Breakwater, there should be a clear gap of 3.5 × Beam (Beam being the width of the largest tanker) between the waiting cargo vessel at the jetty and the passing cargo vessel, so as to avoid any possible collision, ramming and grounding risks. In case of POL, crude and gas tankers, this separation distance should not be less than the 1 psi blast over pressure distance or the Lower flammability limit (LFL), whichever is greater.

17. Remote operated, Quick Release Hooks (QRH) should be provided in all the bollard points so as to release the tanker immediately off the jetty, should there is a fire on board the jetty or on board the tanker. The unit with triple hooks for the head and stern lines should be preferred.

18. Fire Fighting facilities should be checked before ETA of the tanker and kept ready in perfect working condition before a tanker works at the jetty, otherwise shut down should be taken. There should be no compromise on this issue.

19. If the jetty is situated on the leeside of a breakwater section, the transmitted waves through the section may push away the tanker off the jetty face. Secondly, there should be no overtopping of waves

at the jetty site. If required crest elevation may have to be suitably increased to avoid the overtopping of the waves and the section may be made impervious.

20. All the three points, i.e., 16 to 19 should be ensured even if crude/product tanker are handled in the jetty in addition to LPG/LNG.

References

1. Liquefied Gas Handling Principles on Ships and Terminals – McGuire and White.

2. Port Design – Guidelines and Recommendations – Carl A. Theoresen.

3. Planning and Design of Ports and Marine Terminals – Hans Agerschou and others.

4. Port Engineering – Per Bruun

5. Liquefied Gas Carrier Safety Course, Centre for Advanced Maritime Studies, Edinburgh.

6. Practical Ship Handling – Capt. Malcolm Armstrong.

7. Basic Ship Handling – P.F. Willerton.

8. Design of Ashdod Port Extension – D. Helber, Chief Engineer, Israel Port Authority and others.

9. Project Report including Risk Analysis for the LPG jetty, Gujarat Pipavav Port by United Consultants (author's firm) in association with Triune Ltd., New Delhi (unpublished).

10. British Standard Code of Practice for Marine Structures BS 6349

    (a) Part I – General Criteria

    (b) Part II – Design of Quay walls, Jetties and Dolphins

    (c) Part IV – Design of Fendering and Mooring System.

11. Indian Standard Code of Practice for Planning and design of Ports and Harbours: 4651

    (a) Part III – Loading

    (b) Part V – Layout and Functional Requirements

12. Indian Standard Code of Practice for plain and Reinforced concrete: IS--456

13. Permanent International Association of Navigation Congresses – Report of Working Group-IV

14. Wind, Waves and Maritime Structure – R.R. Minikin

15. Hand Book for Ultimate Strength Design of Reinforced Concrete Members – V.K. Ghanekar, R. Chandra and G. Sarkar – Structural Engg. Research Centre, Roorkee.

16. Company Broucher of Rane Elastomer Processors – Cell fenders, Bombay.

17. Company Broucher of Sibata Fenders, Japan.

18. Technical standards for Port and Harbour Facilities in Japan – Bureau of Ports and Harbours, Ministry of Transport, Tokyo, Japan.

19. Pile Foundation, Analysis and Design – H.G. Poulas and E.H. Davis.

20. Vishakhapatnam Port Master Plan studies – Vol.III, Indian Ports Association, New Delhi – L.N. Patnaik, Project Director, J.E. Sivarama Krishnan, Deputy Project Director and others (unpublished).

Index